Black Hand Gorge

A Journey Through Time

Aaron Keirns
3-30-96

To Bernice, with love.

and

To my mother,
Madalene Rina (Tinon) Warman,
for having the foresight to whack me with her
broom whenever I tried to skip school.
Thanks Mom, it worked!

First Edition 9-95

The information printed herein was obtained from a variety of sources including: old books, newspapers, photographs, and oral interviews. We have attempted to verify all information, but due to the historical nature of the subject matter, accuracy is not always certain. Due to the age of some of the photographs in the book, the name of the photographer is not known.

Printed in the United States of America.

Little River Publishing
P.O. Box 291
Howard, OH 43028

ISBN 0-9647800-0-3 (Soft Cover)
ISBN 0-9647800-1-1 (Hard Cover)

BLACK HAND GORGE
A Journey Through Time
Aaron J. Keirns

Acknowledgements

Researching history is like walking down an unfamiliar path. You never know what you might find around the next turn – that's part of the fun. Researching the history of Black Hand Gorge required a good deal of walking. The path I followed led me out of the gorge and into libraries, museums, antique shops, used book stores, and even the courtroom. Best of all, it led me into the living rooms of some fine folks, who have since become friends. Thank you all for your help and hospitality!

A special thanks to my family for their support and patience during this long project.

Thank you…

Curtis W. "Bud" & Rena Abbott
Robert W. Alrutz
Marvin & Victalia Baker
Casey & Tracy Beckett
Chance & Donna Jean Brockway
Mark & Patty Carro
Guy L. Denny
Janet Doyle
Grace Gault
James & Jackie Gerber
Greg Hewitt
Richard W. & Sararose Hewitt
Wilda Hewitt
Hayden Hubbard
Wilma Hunt
Billy & Barb Keirns
Clyde A. Keirns
Gerald L. Keirns
Don & Marcia McFarland
Warren Mears
Diane M. Rutherford
Chester Sidle
Dale Sidle
Kenny & Evelyn Sidle
Robert L. Simon
Goldie Stevens
Mary Walker
Bob Walrath
June Walrath
Edgar Frank & Madalene Warman
Alice Weaver
William E. "Bill" Weaver

Thanks also to…

Fine Line Graphics
Granville Public Library
Johnson-Humrickhouse Museum
Licking County Historical Society
Newark Public Library
Ohio Department of Natural Resources
Ohio Historical Society
Public Library of Mt. Vernon & Knox County
Toboso Elementary School
Zanesville Public Library

In Commemoration
of the
20th Anniversary
of the
Black Hand Gorge
State Nature Preserve
1975-1995

Front Cover Scene

An old post card showing a view of the interurban tracks curving around behind Black Hand Rock.

Photographer Unknown, ca. 1905-1910

Page iii Scene

View of the cliff and stone towpath, as seen from across the river.

Photograph by the author, August 1993

***Page v Scene** (right)*

View of the cliff and stone towpath, taken from beneath a rock overhang across the river.

Photograph by the author, 1972

Table of Contents

Licking Narrows

Come ye lovers, one and all
Licking Narrows sends the call.
On the echoes, through the glades
Gently through the hills it fades.

Come ye where, the flowers fair
Climb o'er painted walls so rare.
Where the rabbits scurry fast
Through the rocky hollows vast.

Where the trees bend down their heads
O'er the ferns and grassy beds.
Where the rippling waters flow
Catching sun or moonlight glow.

Come ye lovers, one and all,
Heed the Licking Narrows call.

Author Unknown

Preface

About This Book

While gathering material for this book, the question I was most often asked was: "Why are you so interested in Black Hand Gorge?" It's a simple question, but I've never been able to answer it very well – I'm not sure I know the answer.

I first saw Black Hand Gorge as a teenager in the 1960s, my older brother brought me here during one of our many arrowhead-hunting expeditions. The gorge was an abandoned area then, overgrown and unprotected. We explored the old interurban tunnel, Black Hand Rock, the railroad cut and the stone steps. From that day on, I was hooked. The gorge seemed like an ancient ruin that had been lost to the forest, I came back often and brought friends with me. We drove our cars down the old interurban railway bed and through the tunnel, and spent long summer days exploring the cliffs and waterfalls. We camped-out on top of Black Hand Rock and skinny-dipped in the old quarry; the gorge became our refuge.

Our long hair, bell-bottom jeans, and beat-up cars made it clear that we were against the war, the establishment, and the pursuit of money. We sat along the river and played folk songs on acoustic guitars. Our harmonies echoed off the giant concave face of Black Hand Rock. People often gathered along the top of the cliff and applauded our impromptu concerts. It was paradise, no rules and no restrictions, I thought I might just stay at the gorge forever. But forever came sooner than I expected. In 1969 Uncle Sam sent me an invitation I couldn't refuse, it included a free haircut.

I wasn't able to get back to Black Hand Gorge for a few years. During that time, the gorge was becoming a trouble spot. Vandalism, drugs, wild parties, illegal dumping, and other problems were getting out of control. Licking County Sheriff's deputies were spread thin trying to police this no-man's-land. Fortunately, in 1975, much of the gorge was set aside as a State Nature Preserve. Since that time the gorge has been a protected area.

Of course, with protection comes restrictions. We can't do whatever we want in the gorge anymore, there are rules and regulations now. But if the gorge hadn't become a State Nature Preserve, it might have become a housing subdivision or landfill by now. We are fortunate that it has remained relatively intact after all these years.

Black Hand Gorge is a natural and historical treasure; education is the key to its protection. The more people that are aware of the historical value of the gorge, the more likely it is to remain protected. The goal of this book is to educate. Through education, we can help protect the unique environment of the gorge for the enjoyment of future generations.

Author's Note

In Black Hand Gorge, there was a group of prehistoric petroglyphs (rock carvings) destroyed during the building of the Ohio & Erie Canal in 1828. The famous Black Hand was just one of the many symbols in this group. These carvings were located on the north side of the Licking River along the south face of what we now call Black Hand Rock. The Black Hand was described as a hand with its fingers "distended" (spread apart) and had the stub of a wrist attached. Some accounts say that the hand had an elongated finger which pointed to an Indian mound nearby. There was an important flint quarry just a few miles south of Black Hand Gorge, today it's a state park called Flint Ridge. It has been suggested that the Black Hand was a signal to people traveling to the flint pits that all tribes or groups must travel through here in peace. A legend says that the Black Hand was the result of a competition between two Indian braves for the love of the same maiden – which resulted in one brave severing his own hand. In any case, the carving of the hand is reported to have been twice the size of a man's hand. It was black in color and stood out prominently against the surrounding reddish-brown sandstone. It's likely that the black color was the result of natural discoloration.

We don't know what the rest of the carvings in this group looked like, but at Hanover, a small town near the upper end of the gorge, there was another group of carvings referred to as the Newark Track Rock. This group was a little better documented but it too was destroyed, during the widening of State Route 16 in the mid-1960s. There was a drawing of this site made in 1859 which shows that one of the symbols was a human hand. The other symbols were mostly what appear to be "bird tracks" or arrows. Only two small pieces of the Newark Track Rock were saved, one which depicts a face and another the hand. Early

"...'Tis a deep lonely glen, 'tis a wild gloomy place,

Where the waters of Licking so silently lave,

On whose pine-covered summit we hear the deep sigh

When the zephyrs of evening so gently pass by..."

Excerpt from "The Black Hand"
a poem by Hon. Alfred Kelley
Canal Commissioner, 1823-1831

Licking Countians had already defaced much of the Track Rock long before it was destroyed, by carving their own names on top of the ancient carvings.

We know very little about the folks who lived here during prehistoric times. We call these people Indians, Moundbuilders, Adena or Hopewell...but we don't know what they called themselves. They roamed these hills for thousands of years, yet we know them only by the artifacts they left behind. We find only things that didn't deteriorate easily: mounds, bones, stone tools and weapons, copper, pottery, burnt seeds and such. Most of their softer artifacts didn't last, wood, leather, and fabric are seldom preserved for very long in this climate. Even more ephemeral were their beliefs – the things that truly defined their existence. Unfortunately, we can't find these things in burial mounds or village sites. We know where and how these prehistoric people lived, what they ate, how they hunted, how they made their tools, and how they buried their dead. We know the form of their lives, but not the substance.

We don't know why they created great geometric earthworks a few miles up-river at Newark. The Octagon and Great Circle Mounds located there are some of the most spectacular mounds in North America. The scale and symmetry of the geometric shapes are impressive even by today's standards. It's difficult for us to understand why a group of people living at a subsistence level would devote so much work and time toward such enormous projects. We haven't found a rational explanation for these earthen sculptures, many of them don't appear to be fortifications, burials, or living areas. We can only assign the mounds to the catch-all category of "ceremonial."

Popular history says that a Mr. Sherwood was the first European or "white" man to report seeing the famous Black Hand carving in 1816, probably while hunting. But there were other white men who almost certainly saw the gorge earlier. The first to venture into this area was Christopher Gist around 1760. Chaplain Jones and David Duncan passed through here in 1772, followed by William Dragoo in 1786. Phillip Barrack (or Barrick) and his wife were the first permanent white settlers at the gorge. About 1801 or 1802 they built a crude log house near the Licking Narrows. It was made of round logs and had only three sides, the fourth being left open for the fire. Their daughter, Millie, may have been the first white child born on the Licking River.

In an article about Black Hand Gorge, from the November 1906 issue of *Ohio Magazine*, Clement Luther Martzolff wrote: *"It is only in wild, rugged and picturesque regions that legends are born and permitted to live"*. In Black Hand Gorge, history and legend live side by side. We don't know all the facts, so we must speculate, and that requires thinking and imagining. In the process, we put a little bit of ourselves into the story. What we end up with may be history, or legend, but it's probably a little of both.

The Legend of the Black Hand

Many moons ago, long before the pale face came across the Great Water to this land, here upon the bank of the Licking, was the lodge of the great chief Powkongah, whose daughter Ahyomah was fair as the dawn and graceful as the swan that floats on the lake. Her eyes were soft and shy as the eyes of a young deer, her voice sweet and low as the note of the cooing dove. Two braves were there who looked upon her with eyes of love, and each was fain to lead her from the lodge of her father, that she might bring light and joy and contentment to his own. At last said the chief, her father, "No longer shall ye contend for the hand of Ahyomah, my daughter. Go ye now forth upon the war path, and when three moons have passed see that ye come hither once more, and then I swear by the Great Spirit that to him who shall carry at his belt the greatest number of scalps shall be given the hand of Ahyomah, my daughter."

Three months had waxed greater and grown less ere the warriors returned. Then, upon the day appointed, behold, all the tribe gathered to view the counting of the scalps. First stepped forth Wacousta, a grim visaged warrior, who had long parted company with fleet-footed youth, and walked soberly with middle manhood. From his belt he took his trophies, one by one, and laid them at the feet of the chief, while from behind the lodge door Ahyomah, unseen by all, looked fearfully forth upon the scene. With each fresh scalp the clouds settled more and more darkly upon the bright face of Ahyomah, and her lip trembled as she murmured, "So many! so many!" Then came the second brave, Lahkopis. Young was he, with the light of boyhood still lingering in his eyes, but upon his head the eagle feather, telling withal of a strong arm and deeds of bravery. One swift glance he shot towards the lodge of the unseen maiden, then he loosed his belt, and laid it at the feet of Powkongah. Scalp after scalp they counted, while the people bent forward silently, and a little hand drew aside the curtain from the lodge doorway, and a young face looked anxiously yet hopefully forth. Slowly, slowly they laid them down, and at last, behold there was one more, just one more than in the pile of Wacousta.

The young Lahkopis had won! Now strode forth Wacousta, and laid his hand – the strong right hand, that yet had failed to win the prize – laid it upon a rock. Then lifted he his tomahawk high in the air, and with one swift stroke severed the hand at the wrist, and flung it high up against the face of the cliff, saying: "Stay thou there forever as a mark of scorn in the eyes of all men, thou hast let thyself be beaten by the cunning right hand of a boy! Disgraced thou art, and no longer shalt thou be numbered among the members of my frame." And the hand clung to the rock and turned black, and spread and grew until it was as the hand of a giant; and while the chief, Ahyomah and the tribe stood silently watching the wonder, the defeated warrior wrapped his robe about him, spoke no word of farewell, and striding swiftly into the dark depths of the forest, was seen no more by man.

A Journey Through Time

Interurban Tunnel *(left)*
View of the west end of the Electric Interurban Tunnel.
Photograph by the author, May 1995

Black Hand Gorge is a spectacular scenic area nestled among the hills of eastern Licking County, Ohio. In September 1975, a large portion of the gorge became a State Nature Preserve. The Ohio Department of Natural Resources currently owns or manages over 1200 acres here. Carved by the Licking River, the gorge slices through nearly four miles of solid sandstone. The upper end of the gorge begins near Hanover and the old site of Claylick. Claylick was a thriving little town around the turn of the century, but nothing is left of Claylick now except a small cemetery and some scattered foundations among the weeds. Down-river the gorge begins to narrow as the hills crowd in from both sides. In the past, this area was referred to as the Licking Narrows. Trees leaned out over the river from both sides forming a leafy canopy that was often described as "dark and gloomy". Among the hills and ravines of Black Hand Gorge lies a rich history. It's the age-old story of the struggle between man and nature. Man has repeatedly tried to conquer the gorge. Every new form of transportation seemed determined to force its way through this narrow pass. The history of Ohio's early transportation and industry is written in the rocks here.

The builders of the Ohio & Erie Canal were the first to descend upon the gorge. In the 1820's they built a dam here along with three locks, an aqueduct and a stone towpath. Scores of men with picks, shovels, wheelbarrows, black powder, and mules entered the gorge like an army. They dug and blasted everything in their path – Ohio's canal era had begun. Canal boats carried passengers and freight through Black Hand Gorge for many years. But in the early 1850s steel rails began snaking their way through the gorge – the railroad was coming! As the decades passed, the canal traffic began to slow and finally became only a trickle as this powerful new form of transportation took its place. Like the canal builders before them, the railroad men forced their way along the river. The tracks they laid through the gorge became part of a major east-west route. For many years great steam locomotives thundered along the river carrying passengers and freight through the gorge. But the era of steam eventually succumbed to more modern technology. The steam engines were replaced by diesels and, in the 1950s, the tracks were rerouted along a different path.

The steam locomotives were still running when the shrill whistle of another form of transportation was heard approaching the gorge. The amazing interurban railway was being born. Once again an army of men marched into the gorge, digging and blasting their way through. It was the beginning of a new century, and a new way to travel. The interurbans were powered by electricity. They were very convenient and became a popular way to commute for several years. But in the end, they couldn't compete with a new contraption called the automobile. The interurban era lasted a relatively short time compared to the canal era and the era of the steam trains. After the demise of the interurbans, the tracks at Black Hand Gorge were torn out and, for several years, the old rail bed was used as a scenic automobile road. But eventually the road was closed and it too faded into history.

For a number of years quarrying was a thriving industry at Black Hand Gorge. Year after year steam shovels chewed away the hills on the south side of the river. There was also a huge orchard here. Native trees were cut down and their stumps ripped out to make way for thousands of fruit trees planted for the Cherry Hill Orchard. In the early 1900s, oil men drilled wells all around the gorge carving access roads as they went. The hills shook as torpedoes of nitroglycerin were lowered to the bottom of the well shafts and detonated. Even the nitroglycerin factories themselves exploded causing death and destruction here. Probably the most damaging of all was the destruction caused by the Dillon Dam project of the 1940s and 50s, and the widening of State Route 16 in the 1960s. Dozens of homes were torn down, several small towns were partially or completely destroyed, and ancient rock carvings were obliterated.

These repeated assaults on Black Hand Gorge have left deep, permanent scars. The canal locks, the railroad cut, the quarry, and the interurban tunnel are some of the more visible ones. But somehow these scars add to the mystique of the gorge. Over the years, the scars have taken on the soft patina of age. They seem less like scars now and more like ancient ruins. That's part of the allure of this place, you can see history here...and you can feel it. A walk through Black Hand Gorge is like a journey through time.

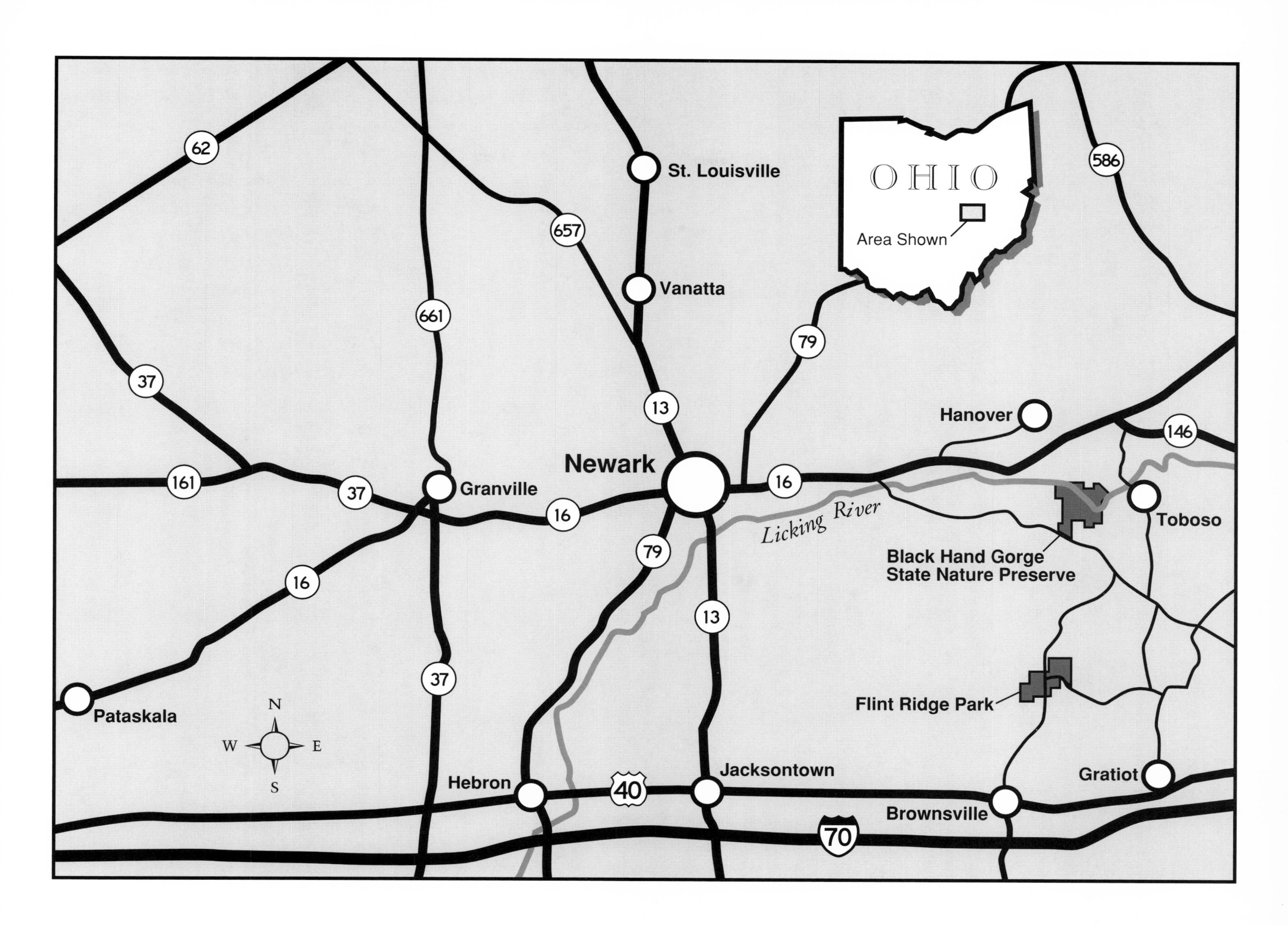

62
St. Louisville
OHIO
Area Shown
586
657
Vanatta
661
79
37
13
Hanover
146
Newark
161
37
Granville
16
16
Toboso
Licking River
79
Black Hand Gorge
State Nature Preserve
16
13
37
Flint Ridge Park
Pataskala
N
W
E
S
Hebron
40
Jacksontown
Gratiot
Brownsville
70

The Canal Era

CHAPTER 1

Chapter 1

The Canal Era

Black Hand Dam *(left)*
The dam became a popular fishing spot for the men and boys of Toboso.
Photographer Unknown, ca. 1895
Collection of Chance Brockway

On July 4th 1825, a large group of politicians, soldiers, musicians, and well-wishers gathered at the Licking Summit near Newark Ohio to witness an historic occasion. Ohio's Governor, Jerimiah Morrow, and Governor Clinton of New York threw out the first spadefuls of earth for what was to become the Ohio & Erie Canal. Like its name suggests, this grand canal system would link the Ohio River to Lake Erie.

At this time in Ohio, there were no railroads and few wagon roads of any worth. Farmers in Ohio's fertile valleys grew fine crops but their markets were almost inaccessible. Surpluses were high and prices for goods were severely depressed. Ham brought only 3¢ per pound, eggs 4¢ per dozen, flour $1 per hundred pounds, and whiskey 12 1/2¢ per gallon. The canals would change all that by linking the state's farmers with markets fed by Lake Erie and the Ohio River. The building of the canals also stimulated the state's economy by providing work for Ohioans, many of whom were the farmers themselves. Typical wages for laborers building the canal were $8 for 26 working days or about 30¢ per day. A day's work lasted from sunrise to sunset. It was hard work but meals and lodging were provided and the opportunity to make cash money was a strong incentive. Small measures of whiskey called "jiggers" were provided to each man, usually at 10 a.m., 12 noon, 4 p.m., and again before supper. This curious practice led to problems and was discontinued by the canal commissioners after only a few months into the project.

The canals provided a variety of benefits to Ohio. Agriculture and industry benefited because prices for their goods went up and the cost of what

Canal Markers *(above)*
These markers are located in Heath, Ohio along State Route 79. One wall of a canal lock is still standing near the markers.
Photographs by the author, June 1994

Township Map *(far right)*
Detail of a Hanover Township map from the 1875 Atlas of Licking County, Ohio. (Captions added).

they imported went down. Production and development increased across the state, towns and cities grew and prospered along the canals, and Ohioans were able to travel more often and for longer distances. The canal system contributed greatly to the unity and prosperity of Ohio's early years.

The specifications for the construction of the Ohio & Erie Canal called for the canal to be no less than 40 ft. wide at the top and 26 ft. wide at the bottom of the channel. It was to have sloping sides and a minimum depth of 4 ft. The canal usually ran alongside rivers. On one of the banks, usually the river side, was to be a towpath 10 ft. in width. For a variety of reasons, these specifications were often exceeded. The Ohio & Erie Canal channel varied in width from 40 to 150 ft. wide at the waterline and was generally 5 to 12 ft. deep. Other features such as locks and aqueducts were more critical to the proper operation of the whole system and were required to meet more exacting specifications. To provide enough water to operate the canal, reservoirs were constructed at strategic locations. At the Licking Summit an area commonly called the "Great Swamp" was dammed-up to create the large reservoir we now call Buckeye Lake.

From the Licking Summit, the path of the canal headed east along the north side of the Licking River toward Black Hand Gorge. Not far from the villages of Claylick and Hanover, the canal had to cross the Rocky Fork Creek. To carry the canal boats over the creek, a stone aqueduct was built. An aqueduct was essentially a bridge designed to carry water over water. This particular aqueduct is made of blocks of Black Hand Sandstone quarried nearby. Its two graceful arches are still standing today – a tribute to the skilled craftsmen who built it.

It's been more than a hundred years since the last canal boat floated lazily across the Rocky Fork aqueduct, yet this relic of bygone days is still carrying farm goods and people across the creek here. Its clay-lined channel is no longer filled with canal water, but hard-packed earth instead. The aqueduct now serves as a bridge on a private farm road. The barges and passenger packets of the 19th century, have been replaced by the tractors and pickup trucks of the 20th century.

Just south of the Rocky Fork Aqueduct, a "lock" was built. Locks functioned as elevators for boats. They were the mechanism that enabled the canal to adjust itself to the lay of the land. Locks on the Ohio & Erie Canal were built of stone blocks and had a set of large wooden gates at each end. A resident lock tender would open and close the gates to allow boats going in either direction to be raised or lowered as needed. Many years ago, a house stood directly on top of this abandoned lock and the stone lock chamber served as its basement. Unfortunately, both house and lock are now gone.

The huge rock formations in the gorge presented a great obstacle for canal engineers. There was no way around the enormous ridges and cliffs there. The only practical way to get the canal through the gorge

What's the Number?

Canal locks are identified by numbers referring to their position in the canal system. In Black Hand Gorge, lock numbers refer to the lock's position relative to the Licking Summit (Buckeye Lake). For example, Lock #15 would be the 15th lock in sequence from the Licking Summit. There has been some confusion about the numbering of the locks in the gorge. For example, in Jack Gieck's book, "A Photo Album of Ohio's Canal Era, 1825-1913", the outlet lock at the upper (west) end of the gorge is identified as Lock #15. A profile map of the Ohio & Erie Canal, published by the Roscoe Village Foundation, also shows this lock as #15. In a report entitled "Cultural Resource Assessment of the Blackhand Gorge Nature Preserve", published by the Licking County Archaeology & Landmarks Society, a map shows this same lock as #17. However, this report also indicates that the lock was referred to as #15 in the earliest records. Based on this evidence, it's seems likely that the missing lock near the aqueduct is probably #14, the outlet lock is #15 and the guard lock #16, but determining which set of numbers is correct is beyond the scope of this book. For our purposes, we will refer to the two existing locks simply as the upper lock and the lower lock.

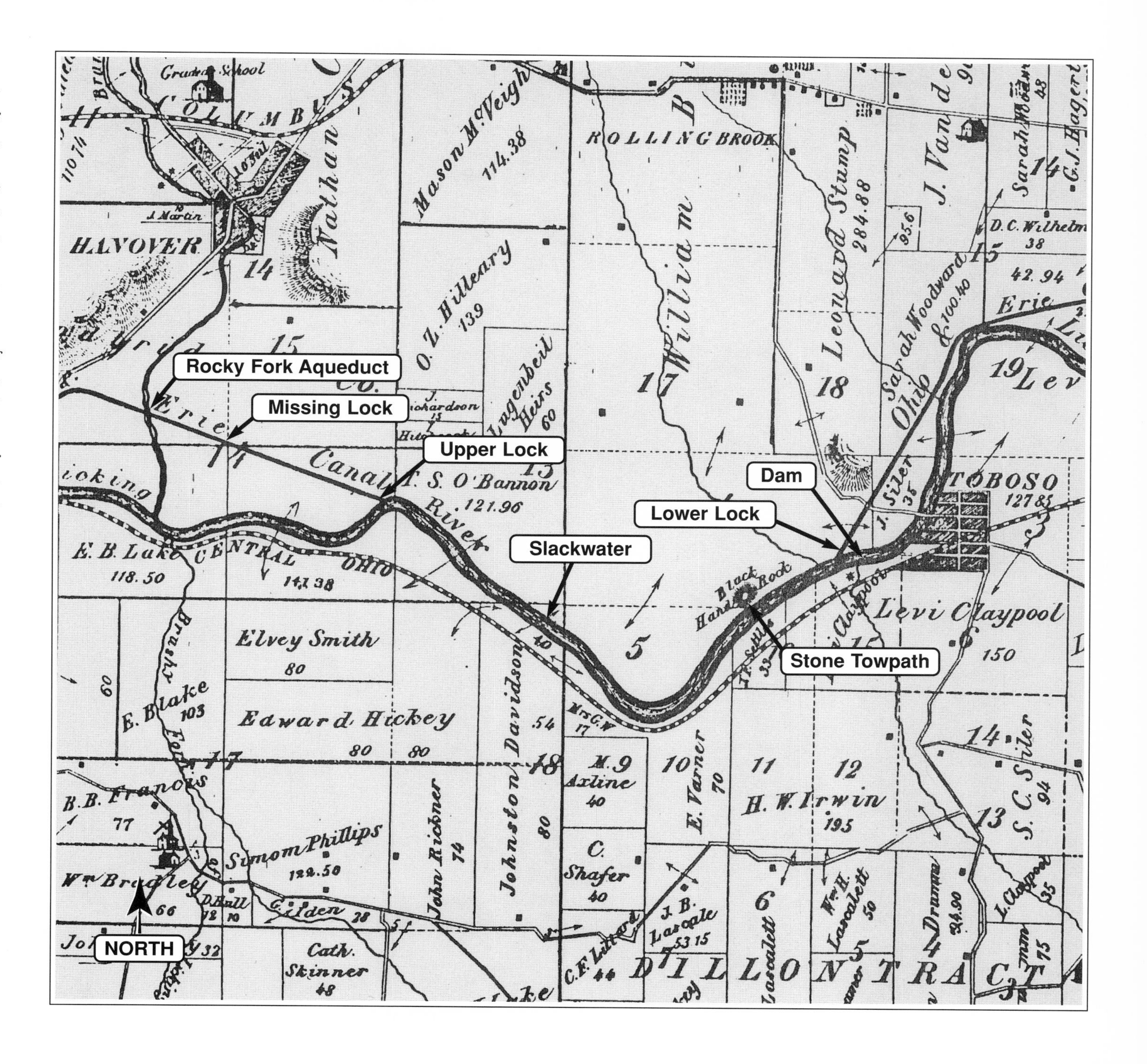

Rocky Fork Aqueduct *(far right)*
This kind of aqueduct, with its stone arches and clay-lined channel, is referred to as a culvert. Stone culverts were built across small creeks while wooden aqueducts were used to span larger rivers. There were exceptions, however, and some stone culverts were longer than some wooden aqueducts. Wooden aqueducts were notorious for springing leaks and required seasonal repair to replace rotting wood and rusting bolts. Stone culverts seldom leaked and required little maintenance.
Photograph by the author, April 1994

Canal Laborers *(below)*
The glamorous life of canal workers.
Courtesy of the Roscoe Village Foundation

was to run it right in the Licking River. In order to use the river as a canal, a "slackwater" had to be created. The water level needed to be raised and the current slowed down. Pulled along by mules or horses, canal boats had no power of their own and they weren't designed to navigate in fast moving water or rapids. To create the slackwater, a dam was built across the Licking River at the lower end of the gorge. It was located about 100 yards upriver from the current road bridge at Toboso. The dam was made of large timbers which were wedged into vertical grooves carved into the rocks on the south side of the river. On the north side, the dam was anchored by a stone abutment. The grooves and abutment are still visible today. The dam created a small lake at the base of Black Hand Rock.

At the west (upper) end of the narrows, a lock called an "outlet lock" was built where the canal connected to the river. This lock served to raise or lower the canal boats to the level of the slackwater, which might vary depending on the level of the river. Another lock was built at the east (lower) end of the slackwater where the canal channel met the river. This lock was called a "guard lock" and functioned to guard the canal from any high water that might come from upriver. The guard lock can be visited along the Canal Lock Trail. Since these two locks functioned as gates in and out of the slackwater impoundment, they had no specific amount of "lift", the lift here could vary depending on the level of the river. The maximum lift of the east lock was about eight feet.

We are fortunate that several important pieces of the Ohio & Erie Canal still stand in Black Hand Gorge. The Rocky Fork Aqueduct, the outlet lock, the stone towpath, and the guard lock are all masterpieces of stone masonry. Built of native materials by common men with simple tools, they have weathered nearly two centuries of use and neglect. These relics of the canal era seem randomly scattered across the landscape now, the ribbon of water that once connected them is gone. Anachronisms in this age of super-highways and air travel, they were once important links in a grand system that transported people and goods from the Ohio River to Lake Erie.

The Rocky Fork Aqueduct

Up-stream View *(above)*

The aqueduct as viewed from up-stream on the Rocky Fork Creek.

Reflections *(right)*

The graceful symmetry of the arch is reflected in the still water beneath the aqueduct.

Photographs by the author, Spring 1994

Between the Arches *(above)*
The roots of trees and other plants are invading the stonework.

Canal Channel *(left)*
Frank Warman and Nathan Keirns are shown standing in the old canal channel on top of the aqueduct.
Photographs by the author, June 1994

The Upper Lock

The upper canal lock is still in fairly good shape but there has been some damage to the west end from vandalism or natural causes. Several stones have fallen out and others are in danger of falling. Due to its low elevation, there is often a foot or two of water standing in the lock chamber. Like many old canal relics, this wonderful old lock is suffering from neglect. It's still relatively intact but, if it continues to be neglected, it may be beyond repair in a few more years. The upper lock is located in a restricted area of the Black Hand Gorge State Nature Preserve. It may be visited only with the permission of the Ohio Department of Natural Resources.

Looking West *(right)*
This photograph was taken just before dark on a warm summer evening. A recent dry spell had left the lock chamber almost dry.
Photograph by the author, August 1993

Looking East *(far right)*
This view, looking east toward the river, shows the stone blocks which have fallen into the channel at the west end of the upper lock.
Photograph by the author, April 1994

The Lower Lock

The lower lock is located on a walking path and can be visited by following the Canal Lock Trail which is located just across the river from Toboso. The old canal channel can be seen leading up to the lock from the river. Years ago, a gravel quarry was located near the lock and the lock chamber was used as a convenient place to load trucks. During the spring of 1986, archaeologists from the Ohio State University conducted an archaeological investigation of Black Hand Gorge. Excavations in the lock chamber revealed one of the original butterfly wicket valves buried beneath the silt. The wicket valves were essentially small iron doors located near the bottom of the large wooden gates and were used to fill and drain the lock chamber.

Lock in Winter *(right)*
Looking northeast into the lock chamber from the channel leading to the river. The sloping walls of the canal channel stand out clearly against the dark stones on this snowy day. The channel was sometimes called a "prism" because of the trapezoidal shape of its cross-section.
Photograph by the author, February 1993

Canawlers Waiting *(below)*
Folks who spent their lives on the canals were often referred to as "canawlers". Here several canal boats are shown waiting their turn to pass through a lock.
Courtesy of the Roscoe Village Foundation

Lock Wall *(above)*

The recessed areas in the lock wall allowed the large wooden gates to open flush with the wall so as not to obstruct the passage of the canal boats. The boats were only about one foot narrower that the width of the lock chamber.

Photograph by the author, February 1993

Iron Hardware *(right)*

The axis on which the heavy lock gate pivoted was a large wooden post called the heel post. The curved "gooseneck" straps shown in this photograph were part of the iron hardware that held the heel post in place.

Photograph by the author, February 1993

The Stone Towpath

The old stone towpath running along the base of Black Hand Rock is a unique relic from the canal era. When the builders of the canal surveyed the "narrows" of the gorge they found that Black Hand Rock jutted out into the river on the north side, leaving no room for a towpath. To make room, it was necessary to blast away some of the overhanging rock. Unfortunately this happened to be near the the part of the cliff where the black hand was carved. The canal workers apparently didn't intend to destroy the black hand carving, they misjudged the amount of black powder needed for the blast. When the powder was set off and the smoke had cleared, the mysterious black hand was gone forever. The mystique of the black hand seemed to intensify after it was destroyed. Many of the canal workers were superstitious, some feared that they would suffer bad luck for their part in destroying this ancient symbol. Boatmen feared to pass between the grim walls of the gorge and spooky stories circulated about this dark and gloomy place. One canal man claimed to have seen "the Devil himself" while passing through the gorge.

After the rock was blasted away, a stone towpath was laid-up along the bottom of the cliff. It's made of rectangular blocks of Black Hand Sandstone which were probably quarried just across the river. The top of the path, where the mules walked, is about 10 feet wide. The towpath is still intact today, although a stone or two is missing and tree roots have damaged the outer wall in some places.

Mule Path *(top right)*
Looking west along the old stone towpath at the base of Black Hand Rock. Bernice and Nathan Keirns are shown walking where the mules would have walked when towing a canal boat through the slackwater of the gorge.
Photograph by the author, February 1993

Water Level *(bottom right)*
The water level was much higher during the canal days due to the dam down-river.
Photograph by the author, March 1993

Scenic View *(far right)*
View of towpath and Black Hand Rock taken from across the Licking River. Notice the two people sitting on top of the cliff.
Photograph by the author, March 1993

Iron Railing, ca. 1905 *(right)*

These two well-dressed gentlemen are standing on the stone towpath at the base of Black Hand Rock looking out over the Licking River. The iron railing was installed several years after the demise of the canal. It was apparently put up to keep visitors from falling into the river during the interurban railway years. The interurban tracks are visible in the distance at the right side of the photograph.

Photograph by J. T. Haynes, ca. 1905
Collection of Chance Brockway

Iron Railing, 1994 *(below)*

The railing is gone now except for a few remnants like this one at the east end of the towpath.

Photograph by the author, November 1994

"... To form the safe towing-path, long since that day

The face of the rock had been blasted away.

Now the gay painted boat glides so smoothly along,

Its deck crowned with beauty and cheerful with song.

And the print of the black hand no longer is seen,

But the pine-covered summit is still evergreen,

And still through the branches we hear the deep sigh

Of the spirits of air as they sadly pass by..."

Excerpt from "The Black Hand" a poem by Hon. Alfred Kelley Canal Commissioner, 1823-1831

Looking East *(right)*
This is the view the mules would have seen when approaching the stone towpath heading east through the gorge (minus the trees).

Towpath Surface *(far right)*
Iron cleats help hold the outer row of stones in place.

Photographs by the author, March 1993

***Gorge View, ca. 1895** (far left)*

This great old photograph shows an unidentified man boating in Black Hand "Lake" sometime before the dam was washed out. Another boat and two more people are visible at the right side of the picture. Notice the high level of the water along the stone towpath at the bottom of Black Hand Rock.

Photographer Unknown
Collection of Chance Brockway

***Gorge View, ca. 1910** (left)*

Compare the water level in this photograph with the earlier photograph at the far left. This later photograph shows that the water level dropped considerably after the dam was destroyed. About halfway up the stone towpath a light-colored row of stones indicates where the water level had been during the canal days. Notice the utility poles which were anchored to the old towpath sometime after the demise of the dam. In the distance, a railroad water tower and the town of Toboso can be seen. The large white building is the Toboso Methodist Church. At the lower right, an electric pole for the interurban railway is visible. The arm sticking out from the pole holds the overhead wire for the electric cars that ran on the tracks at the bottom left corner of the photograph.

Photograph by J. T. Haynes
Collection of Chance Brockway

The Dam

The dam at Black Hand Gorge was made of large timbers and was built around 1830. On the Toboso side of the river, the dam was held in place by the natural rock formations. The other end was anchored by a stone abutment made of rectangular blocks of Black Hand Sandstone. The dam was 250 feet long and had a fall of 17-1/2 feet. In 1831 the first canal boat passed through the locks and slackwater of Black Hand Gorge. For many years, a steady stream of passenger packets and freighters floated through this narrow pass. The last canal boat passed through in 1893, marking the end of 62 years of canal traffic through the gorge. The dam was eventually torn away by a flood in 1898 after holding back the waters of the Licking for nearly 70 years and out-lasting the canal system for which it was built.

***Grooves in the Rocks** (right)*

On the south side of the Licking River, two vertical grooves can be seen in the large rocks along the bank. These grooves were apparently carved to help anchor the end of the dam. This picture was taken from across the river.

Photograph by the author, February 1993

***Old Dam Abutment** (below)*

Between the lower canal lock and the river, remnants of the stone abutment which held the dam in place are still visible.

Photograph by the author, February 1993

End of an Era

***Boating on the Canal** (top)*

The J. T. Haynes family prepares to go boating between 5th and 6th streets in Newark, Ohio in 1898. The young boy in the boat holds on to a puppy while his father prepares to shove off.

Photographer Unknown, 1898
Courtesy of the Roscoe Village Foundation

***Skating on the Canal** (right)*

In areas where the canals weren't drained during the winter, they were a great place to skate. This scene was photographed about 1905 near 6th Street in Newark, Ohio.

Photographer Unknown, ca. 1905-06
Collection of Chance Brockway

By 1900, much of Ohio's canal system had already fallen into disrepair. Some segments of the canals were still in operation, but traffic was light. The speed and efficiency of the railroads had made the canals obsolete. Ironically, during the building of the railroads, canal boats were often used to transport the railroad ties. The great flood of 1913 destroyed much of what remained of Ohio's canal systems. Many locks and canal channels were heavily damaged or washed out by this disastrous flood. Tangled masses of trees and other debris jammed up against locks and bridges. Some locks had to be dynamited in order to remove the obstructions. Due to the widespread damage from this flood, 1913 is generally considered the end of Ohio's canal era.

A Canal Trip Through Black Hand Gorge

To illustrate how the various parts of the canal functioned, let's climb aboard a passenger packet travelling from Newark to Roscoe in the summer of 1860.

Our boat is being pulled along at a leisurely pace of 3 to 4 miles per hour by three mules hitched in tandem. The team is attached to the boat by a tow rope that stretches out almost 200 ft. ahead of us. The mules are followed by a young boy sometimes referred to as a mule skinner or "hoggee" who drives the team of mules mile after mile along the dusty towpath running along the south edge of the canal. A few miles out we begin to pass below the dramatic sandstone cliffs near Hanover and the canal gradually turns right toward the southeast. It's not long before we see the Rocky Fork Aqueduct up ahead and we know we will soon be approaching the "narrows" of Black Hand Gorge. The aqueduct is wide enough to allow for the towpath as well as the canal channel. After crossing the aqueduct, we soon come to a lock. The land slopes toward the Licking River here and a lock is necessary to take the boat down to a lower level. We can see that the large wooden gates are open on our end (the high water end) and closed on the far end (low water end) of the lock.

As we approach the lock chamber, the tow rope is unhooked from the deadeye and long pike poles are used to help guide the boat into the opening. Crew members also help break the boat's forward momentum by using heavy ropes wound around "snubbing posts" at the edge of the lock chamber. The hoggee walks the mules to the other end of the lock to wait as our boat gradually comes to a full stop inside the lock. The lock tender swings the balance beams and closes the large gates behind us. To lower the boat, the sluice gates (also called the paddles or wickets) are opened on the low end of the lock and the impounded water begins to gush out. Our boat gradually drops in the lock chamber. When the water in the lock reaches the lower level, the large gates on the low end of the lock are opened wide, the boat is re-hitched to the mules, and we maneuver out of the lock a few feet lower than we went in. With the mules out front again, we head for the next lock which is located at the point where the canal meets the Licking River. Here we repeat the same "locking through" process as before but as we exit the lock, we will enter the river. Since the canal meets the river here, the towpath can no longer continue on the south side of the canal and the driver and his mules must cross the canal on a "change bridge" designed for this purpose.

For the next two miles the towpath runs along the north edge of the river and the river serves as the canal channel. The river has become very wide in the gorge due to the dam down-river near Toboso. As we are pulled along through the wide slackwater of the gorge, the men push-off against the river bottom with pike poles whenever we begin to drift off course. Suddenly, the peaceful afternoon is shattered by the whistle of a steam locomotive as a train rumbles by us along the opposite side of the river. The startled mules

Overload! *(right)*

A passenger packet loaded for a Sunday School outing at Orange, Ohio in 1885.

Photographer Unknown, July 4, 1885
Courtesy of the Roscoe Village Foundation

Unfortunately, there are no known photographs of a canal boat going through Black Hand Gorge, but there is a story told by Chalmers Pancoast about a canal trip from Newark to the Black Hand Dam in the 1880s. Miss Laura Jones, a high school teacher, took her students on a canal boat ride to the gorge so they could experience its scenic beauty. The boys discovered they could leap up onto the bridges and jump back down on the boat again as it came out the other side. They did this on the Third, Second, and First Street bridges, yelling like a band of Indians. Miss Jones became so frantic over these wild escapades of bridge-jumping, that she never again attempted to treat her scholars to an excursion on the canal. Eventually, she opened "Miss Jones Select School" for charming young ladies.

From Pancoast's book entitled "Our Home Town Memories" (Vol. 1, 1958).

yank our boat toward the bank, kicking and snorting their displeasure. The train engineer looks down on us from his lofty perch, smugly tipping his cap to the ladies on deck. Our mode of transportation suddenly seems very slow compared to the speed of the smoke-belching steam train. In a few minutes it's quiet again.

Almost two miles later we pass Black Hand Rock at the lower end of the gorge. At the base of this cliff the towpath has been laid-up with large sandstone blocks quarried nearby. As the mules begin to cross the stone towpath, the clopping sound of their hooves echoes off the concave surface of the cliff. The young hoggee follows close behind and smiles as his bare feet touch the smooth, cool surface of the stone blocks. We can see the dam just ahead. A few yards above the dam a channel heads out of the north side of the river toward another lock. Several boys from Toboso are fishing on the dam and wave to us as we bear left toward the channel. At this lock we will be raising the boat up from the level of the river to the level of the canal channel. It's the same process as before except this time the sluice gates will let water in from the high side of the lock.

As we exit this lock we gradually pull away from the spectacular scenery of Black Hand Gorge. It's a warm, sunny afternoon and a perfect time to stretch out on top of the cabin and take a long nap as we float lazily on toward Roscoe.

The Coming of the Railroad

Chapter 2

The Coming of the Railroad

Canal travel had just taken hold in Ohio when the railroads were introduced. The railroads were originally intended to serve a subordinate role to the canals, established primarily to provide a link between the various canal systems. Ohio's main canals ran north and south across the state, connecting the Ohio River to Lake Erie. Many of Ohio's most profitable railroads ran east and west, connecting areas of the state that weren't linked by canal. Because of the canal-oriented thinking of the time, railroad construction was plagued by legislation and funding delays.

Some early Ohio railroads offered passengers a choice of steam- or horse-power, with fares that varied accordingly. Early tracks consisted of wooden rails to which a narrow strip of strap-iron was attached.

In 1850, the Central Ohio Railroad Company began laying track through Black Hand Gorge. Like the canal builders before them, the railroaders found that the only way through the gorge was to stay close to the river. From Claylick to Toboso the tracks seldom strayed more than a few yards from the south edge of the river.

Near the east end of the gorge, across the river from Black Hand Rock, the railroad faced a formidable obstacle. Directly in the path of the tracks stood a wall of solid rock 64 feet high and 700 feet thick. There was no way around it. On July 4th, 1850, with picks, shovels, and blasting powder, workmen began forcing their way through the huge obstruction. It was a grueling task, and an outbreak of cholera resulted in several deaths, but work continued. Twelve months later, after many kegs of blasting powder (and whiskey) were expended, the massive "Deep Cut" at Black Hand Gorge was completed.

Deep Cut *(left)*
Steam train chugging east through the deep cut toward Toboso.
Photographer Unknown, ca. 1900
Collection of Chance Brockway

The Deep Cut

The cut we see today has a rugged appearance that looks almost natural in these surroundings. The rocks don't look artificially broken or show many signs of having been blasted. Time has softened their appearance and hidden the scars. For many years, steam engines pulled long trains of freight and passenger cars through the cut. Smoke bellowed out of their stacks in clouds that drifted out over the river. The sounds of the engines echoed inside the narrow walls of the cut, rolling out across the gorge like thunder.

Trains no longer rumble through the Deep Cut. During the construction of Dillon Dam in the 1940s and 50s, the tracks had to be re-routed along higher ground. Trains now cut across the Licking River on a

Deep Cut *(right)*
Two ladies perched on the rocks at the west end of the Deep Cut.
Photographer, Bierberg, ca. 1905
Collection of Curtis "Bud" Abbott

Passenger Train *(far right)*
This train is shown heading east out of the Deep Cut toward Toboso. Much of the hill behind the train is gone now, due to many years of quarrying at the Everett Sand Quarry. The smoke rising from behind the hill in this photograph is probably coming from the quarry. Bicycles roll along this path today.
Photographer Unknown, ca. 1900
Collection of Chance Brockway

high trestle just up-river from the Deep Cut. Today visitors to the Black Hand Gorge State Nature Preserve can walk or ride bicycles through the cut on a paved bike path. The path, which was built directly on the old railroad bed, is officially called the "North Central Bikeway" and is the first section of a planned 110 mile bikeway that will extend from Mt. Gilead southeast to McConnelsville when completed. This scenic 4-1/2 mile section of the bike path which runs through Black Hand Gorge, follows the Licking River from Toboso to the old site of Claylick.

Trestle *(top)*
A visitor walks along the bike path where steam trains once traveled. High above is the trestle that carries trains across the Licking River today.
Photograph by the author, May 1995

Deep Cut Wreck *(bottom)*
Mangled parts of a B. & O. freight train litter both sides of the track at the west end of the Deep Cut. Notice the ladder on the side of the hill.
Photographer and Date Unknown
Collection of Goldie Stevens

Passenger Train *(far right)*
A steam engine pulling a tender and five cars heads into the west edge of Toboso after having passed through the Deep Cut.
Photographer Unknown, ca. 1900
Collection of Chance Brockway

Train Wrecks

Wreck of 1944 *(below)*
Derailed freight cars along the river. The car upside down in the water was hauling whiskey.

Point of Impact *(top right)*
Steam shoots out of one of the engines just minutes after the wreck. Downed wires are visible at the left side of the photograph.

On the Pond Side *(bottom right)*
A young Jim Walcut of Toboso is shown walking near the wreckage at Norman's pond.

Tender *(far right)*
Another view of the tender wedged between the engines. The tender carried a supply of coal and water for the steam engine.

Photographs by Richard Hewitt, July 1944

The bike path is quiet now, but the serenity of the gorge has been shattered more than once by the chilling metallic sound of trains colliding. Two of the more recent collisions occurred in 1944 and 1946.

On July 25th, 1944, two freight trains were steaming toward Black Hand Gorge from opposite directions. The west-bound train was a double-header on its way to Newark, Ohio from Benwood Junction, West Virginia. Its two engines were linked in tandem with a tender in between, and it was pulling 50 cars. The train heading east was a local, going to Lore City from Newark, and was pulling 23 cars. The two trains slammed into each other head-on at Norman's pond between Toboso and the Deep Cut. Some of the railroad men jumped off just before the collision and, miraculously, no one was killed. The force of the impact separated the double-header's tender from its wheels and wedged it between the two engines, suspended above the tracks.

In the collision, six loaded boxcars plunged down the river bank. One of the boxcars, which was loaded with cases of Four Roses Whiskey, landed upside down in the river. Some say that when railroad crews lifted the boxcar up with a crane, local men attempted to "rescue" the whiskey by catching it in buckets as it drained from the car.

The noise of the collision at Norman's pond could be heard up the hill in Toboso. Richard Hewitt, a teenager at the time, grabbed his camera and ran down to the site of the wreck. When he arrived he saw railroad men walking around the wreckage, some with blood on their faces and clothing. Some of the cars had derailed on the river side of the tracks, others on the pond side. Due to a recent dry spell, the pond was dried-up. Local youngsters exploring the wreck were delighted to find dozens of jars of jams and preserves which had spilled out of the wrecked rail cars.

In the Newark Advocate of July 26, another teenager, Dick Seiter, gave his account of the wreck: *"We were swimming in the river just a little way from the tracks. I was out on the diving board when I happened to hear the trains and looked up. I saw them hit. It was like a movie. ...Steam and smoke was everywhere and the steam made an awful, hissing sound."*

Unknown Wreck *(below)*
This photograph shows several men, probably railroad officials, standing on a wrecked train along the river in Black Hand Gorge. This may be a wreck which occurred in 1913.
Photographer Unknown, ca. 1913
Collection of Curtis "Bud" Abbott

Wreck of 1946 *(far right)*
Onlookers watch as four cranes attempt to drag Engine #4300 up the steep river bank.
Photographer Unknown, August 1946
Collection of Curtis "Bud" Abbott

The wreck of August 7, 1946, occurred in the same spot as the head-on wreck two years earlier. Again, it involved a double-header coming from Benwood Junction. This time the fast moving double-header plowed into the back of a local freight train, crushing its caboose like an eggshell. Five trainmen were injured as they leaped from their posts. In the collision, over 100 yards of track were torn up, and a total of 7 boxcars, plus the engines, were derailed. The twisted wreckage attracted hundreds of spectators over several days while the cleanup took place. People gathered on a sandbar in the middle of the Licking River to watch cranes pull the fallen steam engine No. 4300 up the river bank. Some folks even camped-out on the sandbar so they wouldn't miss anything. It took four huge cranes pulling together to lift the big engine up the bank and back onto the tracks. The cable of one of the cranes snapped under the strain, recoiling with such force that it broke the leg of a workman standing nearby.

There have been several other train wrecks in or near the gorge. Robbins Hunter, in his book, *The Judge Rode a Sorrel Horse*, talks about two train wrecks that happened at the gorge in the late 1800s when he was a boy. The first wreck was a passenger train which was heading east through Black Hand Gorge when a large boulder rolled down on the track directly in front of the engine. When the engine struck the boulder, the impact was so great that the engine was thrown into the Licking River, headed in the opposite direction.

Hunter relates the story of another wreck that took place in September of 1890. This wreck was apparently caused by the telegraph operator at the Black Hand telegraph office who neglected to send an order to an eastbound train to hold on the sidetrack while a westbound train passed. The operator quickly realized his mistake, but it was too late. The trains hit head-on just east of the gorge, killing one engineer and seriously injuring the other.

Railroad Memories

Toboso Station *(top)*

The train station in Toboso was located in the lower part of town. The house behind the station in this photograph stands about where the log cabin at the entrance to the nature preserve is located today.

Photographer and Date Unknown
Collection of Goldie Stevens

Hanover Station *(bottom)*

This view shows Pan Handle #19 passing through Hanover. Notice the high bridge in the distance which is still in use today.

Photographer and Date Unknown
Collection of Chance Brockway

Section Gang *(right)*
An unidentified section gang poses near what may be the South Fork of the Licking River.
Photographer Unknown, ca. 1905-10
Collection of Chance Brockway

Work crews called section gangs were responsible for maintaining a specific section of track. The gangs typically worked 10 hours per day, Monday through Saturday. In an old record book from 1898, the foreman of the section gang for section #3 at Stockport is listed as receiving a salary of $40 per month. The other gang members were paid $1.10 per day for each day worked. These gangs spent hours walking their section, constantly repairing the tracks and trestles, tightening bolts, cutting weeds, shoveling gravel, ditching, fixing embankments, loading and unloading ties, and doing whatever else was needed to keep things safe and in good condition.

Railroad Relics

Deep Cut *(top)*
Snow covers the bike path in this view looking west.
Photograph by the author, March 1993

Bridge Abutments *(bottom left)*
Remains of an old railroad bridge at the upper end of the gorge. This bridge carried trains across the Brushy Fork Creek where it emptied into the Licking River. The bike path circles out around this old bridge, crossing the Brushy Fork at a different point.
Photograph by the author, August 1993

Bridge Structure *(bottom right)*
Surprisingly, the heavy stone walls of the railroad bridge are supported by wooden posts and timbers.
Photograph by the author, August 1993

Pump House *(above)*
A few yards west (up-river) from the paved parking area at the Toboso entrance to the nature preserve, there's a large clearing along the south edge of the bike path. This is the site of a railroad water tower. Steam engines stopped here to replenish their water supply. The tower is gone now, but down over the river bank, a small concrete building still exists. It once housed the pump that pumped water up from the river into the water tower.
Photograph by the author, November 1994

The history of Ohio's railroads is a complicated tangle of false starts, failed ventures, construction delays, financial problems, tragic wrecks, grand successes and dismal failures. The railroad in Black Hand Gorge is no exception. The Central Ohio Railroad originally laid the tracks through the gorge in the early 1850s, but it had constant financial difficulty and was finally overtaken by insolvency. The line was placed into receivership in 1859 and operated in that condition until 1865 when it was purchased by the Baltimore & Ohio Railroad. The B. & O. operated the line until just a few years ago, when it became the property of the CSX.

Trains still run through Black Hand Gorge, these days they are modern diesels. They pass regularly along the south edge of Toboso, but they don't stop here anymore. The tracks no longer follow the river as they once did. Like our modern freeways, today's tracks take the shortest path across the landscape.

The rails of the old steam trains are gone, but their scenic path is still here, still carrying travelers along the Licking River – but not on coal-burning steam trains. The times have changed, today's travelers ride on calorie-burning bicycles, roller blades, and high-tech walking shoes.

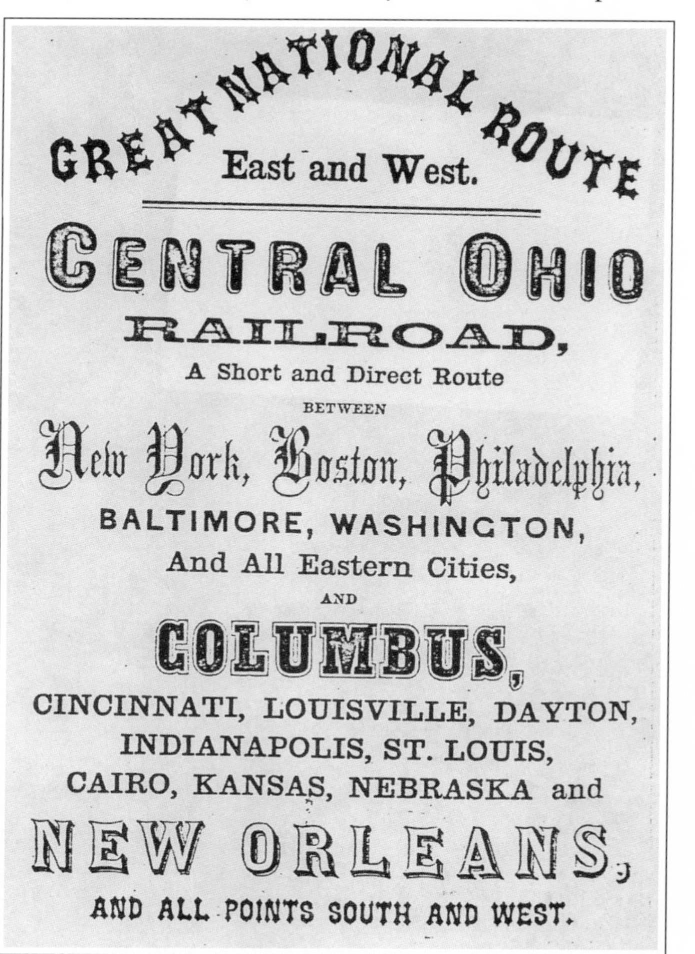

Advertisement *(right)*
An old broadside from the Central Ohio Railroad.
Courtesy of Curtis "Bud" Abbott

Caboose *(far right)*
This large caboose was made from a stock car. It ran on the B. & O. railroad for many years before being retired to the front lawn of a Newark Township residence.
Photograph by the author, October 1994
Courtesy of Curtis "Bud" and Rena Abbott

The Electric Interurban

CHAPTER 3

NEWARK
LOCAL
150
150

Chapter 3

The Electric Interurban

On October 4th, 1890, an electric trolley car traveled from Newark Ohio to nearby Granville, a distance of about 7 miles. Historians believe this may have been America's first electric "interurban" run. Many cities had trolleys that ran within the city limits, but this was the first attempt to connect two towns by trolley. It also became the first electric trolley line to carry the U.S. mail. This novel experiment developed into an entire industry. By the turn of the century, interurbans were springing up all over the state and the nation.

In some ways interurbans were like our buses of today. They could stop at small towns, road crossings, or just about anywhere someone wanted to get on or off. They came at frequent intervals, often running every hour or even half hour, and they were very fast. They could travel at 60 to 80 miles per hour in open country, a blazing speed for those days. Because they were electric, they were clean, quiet, and relatively safe. Interurbans posed a real threat to the steam railroads which ran only two or three times a day and could stop only at stations.

The interurbans changed the lives of many rural Ohioans. Before the interurban, about the only way to get to town was by horse and buggy. The roads were not much more than hardened ruts and often became impassable quagmires after a rain. A trip to a town 30 miles away might be considered an annual excursion. But the interurbans made it possible for rural folks to zip into town, do some shopping, visit friends, and be home in time for supper. The electric interurban craze peaked just before the First World War. It was replaced by another craze whose peak has yet to be reached...the automobile.

Interurban Car *(left)*
The Conductor and Motorman of car No. 150 pose proudly in front of their vehicle. This is one of the cars that ran frequently through Black Hand Gorge.
Photographer Unknown, ca. 1918-20
Collection of Chance Brockway

The Tunnel

***Picnic Rock** (far right)*
Two proper ladies await the arrival of the next interurban car at the east end of the Black Hand Tunnel. The path behind the ladies leads up to the top of Picnic Rock. This path is still visible today.
Children who lived and played near the gorge during the interurban years, sometimes needed to pass through the tunnel to get to the Cornell steps or back to Toboso. The interurban cars were large and fast and there wasn't much room between the tracks and the tunnel walls. The children didn't want to be caught in the tunnel when one of these cars came through. They learned that the least likely time for a car to come was immediately after a car had just passed through. They would wait patiently until a car came and stopped at Picnic Rock, as soon as it departed, they would run as fast as they could through the darkness of the long tunnel.
Photograph by Bierberg, ca. 1905
Collection of Chance Brockway

Like the builders of the canal and the railroad, the builders of the interurban railway were also confronted by a solid rock barrier at Black Hand Gorge. The interurban chose a route along the north side of the Licking River, the south side was already occupied by the B&O Railroad. As the route proceeded east through the gorge, it curved north with the river and then swept around behind Black Hand Rock. Dead ahead lay the sheer walls of Red Rock.

There was no way to get around Red Rock, so the engineers decided to tunnel through it. This was no small undertaking. Red Rock is over 300 feet thick at the point where the tracks were to go through. But the builders of the interurban had one big advantage over the builders of the canal and the railroad – they had dynamite. After dynamite was invented in 1866, blasting became more precise and predictable than with black powder – but it was no less dangerous.

At 3:30 in the afternoon of July 22, 1903, a terrible explosion rocked the unfinished tunnel. Fifty sticks of dynamite had accidentally discharged, badly injuring eight workers. The force of the blast drove huge fragments of rock and showers of sand in every direction, filling the tunnel with dense clouds of smoke and dust. Dr. Postal of Black Hand (Toboso) was soon on the scene but it took two hours for two other doctors to reach the area in their horse-drawn buggies. Two of the workers had parts or all of their hands blown off in the accident, while others sustained injuries such as burns, lacerations, and broken bones. A report in the Semi-Weekly Advocate of July 24th, 1903, stated that:

"The names of the injured could not be learned and would be unintelligible if printed, as they were all Austrians, who can speak hardly a word of English."

The doctors on the scene commented that some of the injuries might turn out to be fatal. The next day, four of the men were laid on cots in the baggage car of a B&O train and taken to St. Anthony's Hospital in Columbus.

It took nearly 3-1/2 months to complete the tunnel, with day and night shifts of 36 men. Work on the tunnel began on June 15, 1903 and was finished September 23rd of that year. It was January 1904, however, before all the work between Newark and Zanesville had been completed and the line could be opened for business.

The interurban tunnel at Black Hand Gorge measures 327 feet long, 19-1/2 feet high, and 16-1/2 feet wide at the base. Its graceful shape and uniform size show the pride and skill of the workmen who built it. Interurban tunnels are not common, some people believe this may be the only one in Ohio, and possibly the nation. Due to its length, the tunnel is dark inside even during the day. Along the ceiling of the tunnel circular indentations can be seen about every 4 or 5 feet. These are the result of the blasting during the building of the tunnel. Look a little closer and you may see some little brown critters hanging around the tunnel – these are the tunnel's resident bats.

PICNIC ROCK

Lantern *(below)*
Lantern with red glass from the Columbus, Newark, and Zanesville Electric Railway (C.N.&Z.). The C.N.&Z. was the predecessor of the Ohio Electric Railway Company which took over the line in 1906.
Photograph by Chance Brockway

News Clipping *(right)*
Article from the Zanesville Signal, July 14, 1913.
Courtesy of the Zanesville Public Library

Tunnel Vision *(far right)*
The tunnel is fairly dark inside even in the daytime. Visitors near the other end appear as silhouettes against the outside light. The best time to view the inside of the tunnel is after a heavy snowfall. A surprising amount of light is reflected into the tunnel on a bright snowy day. At night it's virtually impossible to see anything inside the tunnel. The total darkness is disorienting, there are no visual clues by which to judge distance or direction.
Photograph by the author, January 1994

After the interurban line was completed, Black Hand Gorge became a popular tourist attraction. The interurban railway made it easy to travel to this remote area. Red Rock came to be known as Tunnel Rock or, more often, Picnic Rock. The beautiful scenery of the gorge attracted visitors from Buckeye Lake, Newark, Zanesville, and many points in between. A small platform was built at the east end of the tunnel so picnickers could step off the interurban cars there. A path led from the platform to the picnic area above the tunnel. Some of the high ground north of the tunnel was subdivided into lots and became known as Rock Haven Park. Several cottages and a dance pavilion were built there. Some of these cottages still exist today, many have been remodeled and enlarged by the current owners. What started out as a group of country cottages along the interurban line, has evolved into a small community along Rock Haven Road.

An Ohio Electric interurban car left Buckeye Lake heading for Zanesville at 9:30 p.m. on July 13, 1913. It was a Sunday evening and storm clouds were gathering. The 30 passengers on board car No. 153 had spent the day at the lake and were returning home. By the time they reached Black Hand Gorge, they found themselves in the midst of a violent storm. Visibility was poor in the driving rain. They didn't know that the tracks had been washed out up ahead near Black Hand Rock. The situation could have been disastrous, but Mr. Blackwell, the alert conductor, and Mr. Hager, the motorman, saw the problem in time to bring the car to a stop at the edge of the washout. The car became stalled there, so they decided to abandon it and head for shelter. Luckily a telegraph pole had fallen across the little creek that had caused the washout. The men on board helped the ladies cross over on the pole and the whole group headed east along the tracks in the pouring rain. Eventually they reached the Black Hand Tunnel. It was dark and cold inside, but at least it was dry. They spent the rest of the night in the tunnel trying to dry out while the storm raged outside. By Monday morning they were tired, cold, and hungry. Local farmers came to their rescue with blankets and food and they all eventually got home to Zanesville by other means.

THIRTY COMPELLED TO SPEND NIGHT IN O. E. TUNNEL

Zanesvillians in Party Marooned at Black Hand For Hours

TWO DROVE TO THE CITY IN THE MORNING

Party Returning From Buckeye Lake Had Unpleasant Experience.

About 30 Zanesville people spent Sunday night in the Ohio Electric tunnel near Black Hand, being marooned there because of the big rainfall which caused a washout on the Ohio Electric interurban lines. Adolph Berger of the Union Clothing Co. and Joseph Federman of the Zanesville Dry Goods Co., hired a rig Monday morning at 5 o'clock and drove to Zanesville landing here about 1 o'clock in the afternoon. The rest of the party were still at Black Hand when they left Monday morning and had not yet arrived at Zanesville Monday afternoon at 3 o'clock.

The people were on the Ohio Electric car that left Buckeye Lake at 9:30 o'clock and were in the midst of the big storm. As they approached the tunnel at Black Hand the car was stalled because of the washout at this point. The passengers crowded off the car and with

Interurban Tickets *(top right)*

Old tickets from the Columbus, Newark and Zanesville Electric Railway Company.

Collection of Chance Brockway

Black Hand Station *(right)*

The Black Hand Interurban Station was located at the lower end of the gorge, east of the tunnel. This photograph shows passengers awaiting the arrival of car No. 166. The blurred areas of the picture were caused by movement which took place during the relatively long exposure time of the camera.

Photographer, Unknown, ca. 1915

Collection of Chance Brockway

View of Toboso *(left)*

This photograph was taken from the top of the tunnel looking toward Toboso. The square platform at the bottom of the picture is the interurban stop for Picnic Rock. The water tower for the B&O Railroad can be seen just across the Licking River. The large white building is the Toboso Methodist Church. This print was made from an old glass negative which was cracked across the top.

Photographer, Bierberg, ca. 1905

Collection of Chance Brockway

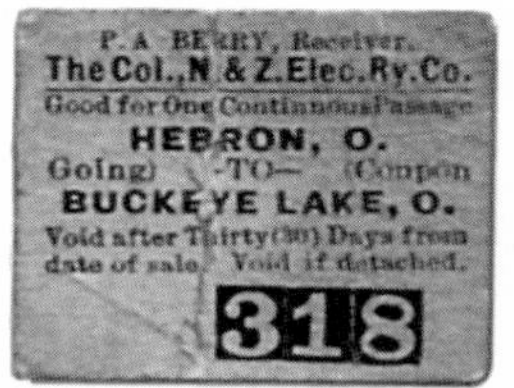

***Trolley & Train** (top right)*
This photograph shows an interurban car heading east through Black Hand Gorge while a passenger train heads west on the other side of the Licking River. The buildings of the Everett Sand Quarry can be seen along the right side of the river. The interurban and the railroad often competed with each other for passengers and light freight.
Photographer Unknown, ca. 1907
Collection of Curtis "Bud" Abbott

***Hanover Station** (bottom left)*
The Ohio Electric Railway station at Hanover near the upper end of Black Hand Gorge.
Photographer Unknown, ca. 1915
Collection of Chance Brockway

***Destination Signs** (bottom right)*
Signs which were mounted in the front windows of the interurbans.
Photograph by Chance Brockway

***West End of Tunnel** (far right)*
The photographer stood on Black Hand Rock to take this view looking down at the tunnel. Notice the camping tents on top of Picnic Rock.
Photographer, Bierberg, ca. 1905
Collection of Chance Brockway

Time Table *(right & below)*
Interurban cars were often referred to as traction cars. This "Time Table of Traction Cars", from April 1927, is printed on light blue paper and unfolds to a size of 7" x 12". On its cover is a view of the "Licking Gorge", better known as Black Hand Gorge. The area between Zanesville and Newark was considered by many people to be the most scenic interurban route in the state. The fare for riding from Zanesville to Newark was 80¢.

Photograph by the author, August 1994
Courtesy of James & Jackie Gerber

The Lookout *(far right)*
A boy in knickers stands on the roof of the Black Hand Interurban Station around 1910. Notice the oil well derrick in the distance and the crates stacked near the station door which were probably used to ship farm produce to town.

Photographer Unknown, ca. 1910
Collection of Curtis "Bud" Abbott

After the demise of the interurban, Hugh Kennedy, who had been the station agent since 1911, purchased the Black Hand station building along with the waiting room, benches, and even the pot-bellied stove. Hugh moved it all across the road and transformed it into a gas station and store. It became a popular gathering spot and a good place to play dominoes, cribbage, pinochle, and even slot machines. Important issues of the day were discussed and settled around Hugh's old stove. He jokingly referred to his little store as the "unofficial recreation and political headquarters for the community". Hugh was quite a character and many people still remember him pulling up his pant leg to prove that his sock was held up with a thumb tack. Of course, local folks weren't surprised by this, everyone knew Hugh had a wooden leg.

EASTERN STANDARD TIME

Read Down — FROM ZANESVILLE TO NEWARK AND COLUMBUS

Fare	Mi	STATIONS	1 am	3 am	5 am	7 am	9 am	11 am	13 am	15 am
0	0	Lv ZANESVILLE	—	—	6 00	7 00	8 00	9 00	10 00	11 00
.25	9	Pleasant Valley	—	—	6 23	7 23	8 25	9 25	10 25	11 25
.35	13	Nashport	—	—	6 30	7 30	8 32	9 32	10 32	11 32
.55	18	Black Hand	—	—	6 38	7 39	8 40	9 40	10 40	11 40
.60	21	Hanover	—	—	6 46	7 47	8 47	9 47	10 47	11 47
.60	22	Clay Lick	—	—	6 48	7 49	8 49	9 49	10 49	11 49
.70	24	Weiant	—	—	6 52	7 53	8 53	9 53	10 53	11 53
.80	28	NEWARK	5 00	6 05	7 05	8 05	9 05	10 05	11 05	12 05
1.15	39	BUCKEYE LAKE	5 30	6 10	7 20	8 20	9 20	10 20	11 20	12 20
1.10	37	Hebron	5 35	6 30	7 30	8 30	9 30	10 30	11 30	12 30
1.30	42	Kirkersville	5 46	6 40	7 40	8 40	9 40	10 40	11 40	12 40
1.40	47	Etna	5 54	6 47	7 47	8 47	9 47	10 47	11 47	12 47
1.60	54	Reynoldsburg	6 08	6 57	7 57	8 57	9 57	10 57	11 57	12 57
1.80	64	Ar. COLUMBUS	6 40	7 25	8 25	9 25	10 25	11 25	12 25	1 25
			am	am	am	am	am	am	pm	pm

STATIONS	17 N	19 pm	21 pm	23 pm	25 pm	27 pm	29 pm	31 pm	33 pm	35 pm	37 pm	39 pm	40 am
Lv ZANESVILLE	12 00	1 00	2 00	3 00	4 00	5 00	6 00	7 00	8 00	9 00	—	11 00	2 10
Pleasant Valley	12 25	1 25	2 25	3 25	4 25	5 25	6 25	7 25	8 25	9 25	—	11 21	Extra
Nashport	12 32	1 32	2 32	3 32	4 32	5 32	6 32	7 32	8 32	9 32	—	11 26	
Black Hand	12 40	1 40	2 40	3 40	4 40	5 40	6 40	7 40	8 40	9 40	—	11 32	
Hanover	12 47	1 47	2 47	3 47	4 47	5 47	6 47	7 47	8 47	9 47	—	11 38	
Clay Lick	12 49	1 49	2 49	3 49	4 49	5 49	6 49	7 49	8 49	9 49	—	11 40	
Weiant	12 53	1 53	2 53	3 53	4 53	5 53	6 53	7 53	8 53	9 53	—	11 44	
NEWARK	1 05	2 05	3 05	4 05	5 05	6 05	7 05	8 05	9 05	10 05	11 05	11 55	3 05
BUCKEYE LAKE	1 20	2 20	3 20	4 20	5 20	6 20	7 20	8 20	9 20	10 20	11 20	—	—
Hebron	1 30	2 30	3 30	4 30	5 30	6 30	7 30	8 30	9 30	10 30	11 30	—	—
Kirkersville	1 40	2 40	3 40	4 40	5 40	6 40	7 40	8 40	9 40	10 40	11 40	—	—
Etna	1 47	2 47	3 47	4 47	5 47	6 47	7 47	8 47	9 47	10 47	11 47	—	—
Reynoldsburg	1 57	2 57	3 57	4 57	5 57	6 57	7 57	8 57	9 57	10 57	11 57	—	—
Ar. COLUMBUS	2 25	3 25	4 25	5 25	6 25	7 25	8 25	9 25	10 25	11 25	12 25	—	—
	pm	pm	pm	pm	pm	pm	pm	pm	pm	pm	am	am	am

Main Line Trains do not run to Buckeye Lake. Passengers for Buckeye Lake transfer at Hebron.

BLACK-HAND

COLUMBUS
LOCAL
243

***O.E. Car No. 243** (left)*
An Ohio Electric interurban car pauses at the west end of the Black Hand tunnel while its crew poses for the camera. The photograph on the right shows the same scene as it appeared about 25 or 30 years later.

Photographer Unknown, ca. 1910
Collection of Curtis "Bud" Abbott

***Automobile Road** (right)*
During the 1930s and 40s, the old interurban rail bed was turned into an automobile road. Known as the Gorge Road, it was used as a short cut to Newark by the people of Toboso and the surrounding areas. Many local folks still have fond memories of riding along this scenic road in the open air of a rumble seat. This road was permanently closed as a result of the Dillon Dam flood control project.

Photographer Unknown, ca. 1930-40
Courtesy, Ohio Dept. of Natural Resources

Tunnel Ceiling *(right)*

Flash photography reveals old iron springs hanging in the darkness of the tunnel ceiling. These are part of the hardware which held the overhead electric wire that powered the interurban cars.

Holes are also evident along the ceiling. During the building of the tunnel, workmen called "Headers" drilled these holes by driving a steel drill into the rock with a 16 lb. sledge hammer. This was known as "driving the head". The headers were followed by the "Benchmen" who drilled similar holes into the lower or "bench" part of the rock. Sticks of dynamite were then inserted into these holes along with blasting caps with wires running to an electric battery outside the tunnel. The charge from the battery detonated the dynamite. After the smoke and dust had cleared, another group called the "Muckers" came in and cleaned up the debris from the blast.

Photograph by the author, June 1994

Inscription *(far right)*

The initials "W. P A." and the date "10-23-38", are inscribed at the west end of the tunnel. This appears to be a reference to the Works Progress Administration, a government jobs program, which may have helped build or maintain the automobile road through the gorge.

Photograph by the author, April 1994

End of the Line

In 1904, Zanesville had only about half a dozen automobiles. Even by 1912 the automobile was still considered to be just a novelty. According to a newspaper article from that year, *"...the auto craze is at its height now and far exceeds the bicycle craze that held the country in its grasp years ago"*. The automobile was seen as another craze and some thought it had reached its peak of popularity. To most people, the new electric interurban system seemed like a safer and more permanent mode of transportation. But as we all know, the automobile was here to stay. Mass production brought costs and prices down, and soon the average family could afford to purchase their own vehicle. Roads and bridges were improved and autos, trucks, and buses began to steal business from the interurban lines. The interurbans never recovered.

The last interurban car rolled through Black Hand Gorge on February 15, 1929, ending nearly a quarter of a century of electric railway travel through the gorge. The Black Hand station buildings were moved across the road and transformed into a gasoline filling station – a sign of the times. Today, there's not much evidence left of the interurban railway at Black Hand Gorge, except for the tunnel. This unique monument to Ohio's interurban era sits almost forgotten now, dark and empty, on a road that leads nowhere.

Interurban Bridge *(right)*
The interurban crossed the Rocky Fork Creek not far from where the Ohio & Erie Canal had crossed many years before. The interurban bridge is gone now but both of the bridge abutments are still standing, just up-stream from the Rocky Fork Aqueduct.
Photograph by the author, April 1994

Tunnel in Winter *(far right)*
A snowy view of the east end of the interurban tunnel.
Photograph by the author, January 1994

Toboso – Hometown of the Gorge

CHAPTER 4

GENUINE OLIVER PLOWS
TOBOSO, O.
POST OFFICE

CHAPTER 4

Toboso – Hometown of the Gorge

The town of Toboso was laid out on land owned by William Stanbery (also spelled Stanberry). With access to the Ohio & Erie Canal and the proposed Central Ohio Railroad, his land seemed like a perfect place for a town. It was surveyed by David Wyrick who would later draw national attention with his discovery of the controversial Newark Holy Stones. Why Stanbery chose the name Toboso is somewhat of a mystery. Toboso sounds like it could be an Indian word, after all, this area is rich in Indian history and legends. But the name Toboso is probably of Spanish origin. In Spain, there is a town named El Toboso. It's mentioned often in the 17th century Spanish novel, "Don Quixote", by Cervantes. Don Quixote, the elderly hero of the story, has become deluded by reading too many chivalric romances and sets out on his horse to fight giants, rescue damsels, and uplift the oppressed. Throughout the story he tries to gain honor and fame in order to win the heart of his lady, Dulcinea of Toboso. A look at the original plat map of Toboso Ohio, reveals that the street now called Third Street was originally named Quixote Street. This leaves little doubt that the name of Toboso was inspired by this famous Spanish novel.

Toboso Post Office *(opposite page)*
A glimpse of the "good old days" in downtown Toboso. This building served as the post office and general store, and later became a gasoline filling station.

Photographer Unknown, ca. 1900
Collection of Chance Brockway

The Founding of Toboso

"Toboso was laid out by the Hon. William Stanberry, but, on account of its sickly location, has always languished."

A curious quotation from the 1875 Atlas of Licking County by L. H. Everts

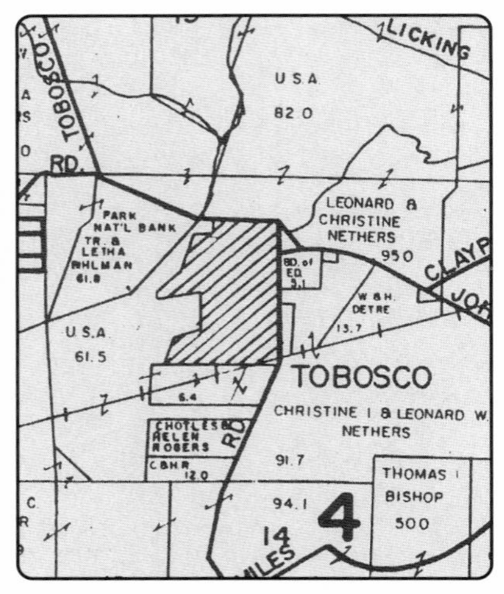

***Map of "Toboso"** (above)*
On some maps, like this example from 1984, Toboso has been misspelled as Tobosco. Consequently, it's often pronounced that way.
©1984 Great Mid-Western Pub. Co., Inc.

In some old books and newspaper articles, the names "Toboso" and "Black Hand" seem to be used interchangeably. It's not always clear whether the writer is referring to the town or the general area. This may explain why it's commonly thought that the town of Toboso was originally named Black Hand, and that the name Toboso was assigned by the U. S. Postal Service when a post office was established in the town. This actually did happen in many small Ohio towns, but not here. Toboso has always been named Toboso. But it's easy to understand why there's confusion. During the canal days, and even earlier, the whole area of the gorge was referred to as Black Hand Gorge, or simply Black Hand. This was before Toboso even existed. After Toboso was founded, many people continued to refer to the entire area as Black Hand, including the town of Toboso. Years later, when the interurban came through, a station was built just across the river from Toboso. Since the station wasn't located within the town limits, it was given the name of Black Hand Station instead of Toboso Station. Anyone riding the interurban to Toboso would actually get off at Black Hand. Another possible reason why the station was named Black Hand was that the gorge was a big attraction for interurban passengers. People rode the interurban from Zanesville, Newark, Buckeye Lake, even Columbus, to visit this scenic area. They came primarily to visit Black Hand Gorge, not Toboso. Also, since the B&O Railroad already had a Toboso Station, it would have created confusion if the interurban station had been given the same name.

The official plat map for the town of Toboso on file at the Licking County Recorder's Office is signed by David Wyrick, L. C. S. (Licking County Surveyor). In this document, Wyrick certifies that *"...the above is a true Plat of the Plan and manner in which the said Town of Toboso is laid out and surveyed by me Dec. 6th, 1831."* The document was witnessed and recorded a few weeks later on January 30th, 1832. There are a couple of problems with this map. The Central Ohio Railroad is shown passing through the center of Toboso, but in 1831 when the map was supposedly drawn by Wyrick, the railroad hadn't even been proposed yet. It wasn't until 1847 that the railroad was chartered and a route decided upon. How could Wyrick have laid out such an accurate map of the railroad right-of-way through Toboso 16 years before the route was known? Records also show that Wyrick wasn't even the Licking County Surveyor in 1831, he didn't become Surveyor until 1850.

This is probably not the original plat map, but a hand-written copy. Whoever copied from the original may have misread the 5s in the dates as 3s, setting the date for the founding of Toboso back 20 years. If the date of the survey was not 1831 but 1851, then everything makes sense. The railroad had been chartered by then and the route for the tracks could have been known by Stanbery and drawn by Wyrick who was indeed the Licking County Surveyor in 1851. It's little mysteries like this that make history interesting!

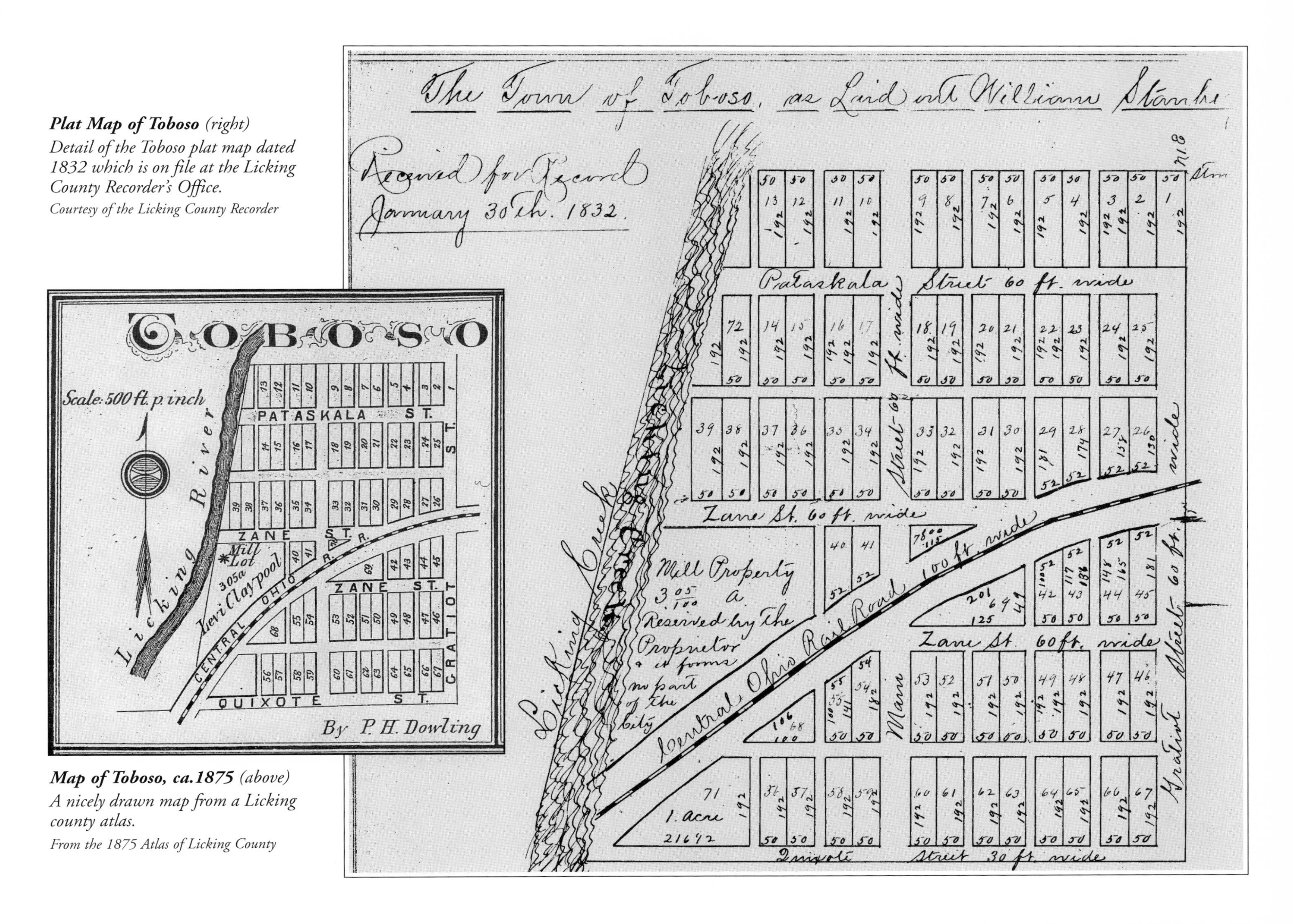

Plat Map of Toboso *(right)*
Detail of the Toboso plat map dated 1832 which is on file at the Licking County Recorder's Office.
Courtesy of the Licking County Recorder

Map of Toboso, ca.1875 *(above)*
A nicely drawn map from a Licking county atlas.
From the 1875 Atlas of Licking County

The Dillon Dam Flood Control Project

Lower Toboso *(left)*
This picture was probably taken from the top of the Toboso school building in the 1920s. The view is looking toward the Licking River. The dark colored building near the upper left part of the picture is the B&O Railroad Station. The building partially hidden from view by the station building is the post office and store pictured at the beginning of this chapter. The white house to the right of the station sits about where the log cabin at the entrance to the Nature Preserve sits today. The wooden structure at the bottom of the picture was used for loading cattle and other stock onto the train cars. An oil derrick can be faintly seen in the distance near the upper right corner of the photograph. All of the houses and other buildings in this picture were moved or torn down due to the Dillon Dam flood control project.

Photographer Unknown, ca. 1920s
Collection of Goldie Stevens

The village of Toboso now occupies only the high ground above the Licking River, but for many years, much of the town was located down on the low ground nearer the river. There were about 20 homes there as well as barns, grocery stores, the post office, and the train station. The Central Ohio Railroad and later the Baltimore & Ohio ran right through the lower part of Toboso. The paved bike path through the Black Hand Gorge State Nature Preserve is built on this old rail bed. The entire lower section of Toboso was torn down in 1959-60 because it fell within the possible flood zone of the new Dillon Dam flood control project. The Dillon project was a controversial topic and met with much resistance from residents of the small towns along the Licking River. They knew that the building of Dillon Dam might cause forced removal from their homes.

The purpose of Dillon Dam was to prevent the Licking River from flooding the rich Muskingum valley. The towns of Zanesville, McConnelsville, and Marietta were all threatened by occasional floods. In 1938, Congress authorized 15 reservoirs to be built to control flooding in the Ohio River Basin, Dillon was the last one of these reservoirs to be built. After work was finally started on the Dillon project it had to be halted temporarily during the Korean War. At Black Hand Gorge, the B&O railroad had to be moved from its path along the river to a more suitable route along higher ground. Massive concrete piers were built in the Licking River to support a new train trestle. The piers sat empty for a few years before the steel bridge sections were finally attached. Resistance from Licking County residents continued, but eventually the project was allowed to resume construction. In a booklet entitled, "Why We Need Dillon Dam", compiled by the Chambers of Commerce of Zanesville, McConnelsville, and Marietta, 15 pages were devoted to explaining the many benefits of the project for towns located below the mouth of the Licking River. Two small paragraphs addressed the concerns of the folks who lived up-river in towns like Toboso. One of these paragraphs began; *"Easements to permit flooding will, of course, also require the removal of houses and many other improvements so that owners will not be inclined to turn out with shotguns to prevent the Government from filling the reservoir in flood times as has happened in some places."*

Residents of small towns along the Licking River in Licking and Muskingum Counties understood the need for flood control, but they also knew that to help save homes down-river, they would have to sacrifice their own. Many small towns above the dam were severely impacted by the Dillon project including Claylick, Hanover, Toboso, White City, Nashport, Irville, and Pleasant Valley. A few homes were moved to higher ground but most were torn down. Some towns, such as Claylick, completely disappeared during this process. Families were uprooted from homes they had lived in for generations, and hometowns were changed forever.

The Toboso Bridge

There have been at least three different bridges across the Licking River at Toboso. A long covered bridge once spanned the river here and served the people of Toboso for many years with a distinct sag at one end. Many people still have memories of this old wooden bridge from their childhood days.

As a young boy in the 1930s, Kenny Sidle remembers leaving Toboso school at lunch time, running barefoot down the hill and across the old covered bridge to Hugh Kennedy's store to get a Tootsie Roll, then scurrying back across the bridge and up the hill to the school before he was missed.

Some people called this the "hugging bridge" because it was a popular place for young couples to hug and kiss without being seen. After many years of service, the old covered bridge was finally torn down in 1941 and replaced with an open iron bridge. The hugging and kissing were apparently moved elsewhere.

The Road to Toboso *(left)*
This view shows the west end of the old covered bridge with the town of Toboso in the background. The photograph was probably taken from the hill just west of Toboso at the intersection of Toboso Road and Rock Haven Road. The small road leading down to the left just before the bridge, went to a group of cottages along the river known as White City. The house near the bottom of the photograph belonged to the Kennedys. Notice the interurban tracks at the bottom right corner.

Photographer Unknown, ca. 1915-20
Collection of Goldie Stevens

Covered Bridge *(right)*
This view shows the bridge as it looked from upstream, probably in the 1920s or early 30s. The pile of boards and debris at the left side of the photograph appears in other old views of the bridge from the early 1900s. In one earlier view it looks like a building or possibly an old abandoned canal boat. Judging by its location, it might be part of the dam which was located here.

Photographer Unknown, ca. 1920s-30s
Collection of Goldie Stevens

Covered Bridge *(top right)*

The same stone pier, supporting the covered bridge in these old photographs, now supports Toboso's modern concrete and steel span.

Photographer Unknown, ca. 1920s
Collection of Goldie Stevens

Later Years *(bottom right)*

Eventually, the covered bridge became unsafe and had to be torn down. While it was being replaced, a temporary foot bridge was suspended across the river on cables. Grace Gault, a former teacher at Toboso, recalls local boys thinking it was great fun to hide in the bushes until someone began crossing the bridge, then they would rush out and swing it back and forth.

Photographer Unknown, ca. 1930s
Collection of Curtis "Bud" Abbott

Iron Bridge *(below)*

The iron bridge which spanned the Licking from 1941 to 1988.

Photographer and Date Unknown
Collection of Goldie Stevens

A Celebration

The iron bridge served the community for more than 45 years, but in 1988 it was torn down to make way for a more modern bridge. Building the new bridge was a big project and took some time. Toboso residents, inconvenienced by the lack of a bridge, anxiously awaited its completion. When it was finally finished they threw a "bridge party" in honor of the occasion. The bridge wasn't open to traffic yet so they held the party right on the bridge. All of Toboso was in attendance. Rows of tables were put end to end and draped with table cloths. It was July 27th and the sun was hot, folks in lawn chairs accumulated in the shady spots. Casseroles, sandwiches, potato salad, baked beans, iced tea, and other home-baked goodies were plentiful. A piano was rolled out onto the bridge and musicians gathered around. The people of Toboso ate, talked, laughed and sang together that day.

A plastic ribbon was stretched across one end of the bridge and Mrs. Wilda Hewitt was given the honor of cutting it, after all, it was her 85th birthday. As the bright pink ribbon fluttered down from Wilda's scissors, the townspeople declared the bridge to be "officially open". Local newspapers reported the story.

Apparently some county officials were not amused by this unofficial official ceremony. They had already planned their own ribbon-cutting ceremony complete with politicians and the press. But now it was too late, the moment had passed. The folks of Toboso had slipped in and taken the bridge right out from under them.

In most small towns, a new bridge would be welcomed – but probably not celebrated with a party. Toboso may have been celebrating more than just a bridge that day, Toboso may have been celebrating itself. This little town, and the gorge, have survived the rise and fall of the canal system, the railroad, the interurban railway, and an oil boom. Each one of these enterprises dug, drilled, or blasted its way through the gorge, bringing with it the promise of new prosperity. Each one, in turn, pulled out and left nothing behind but scars on the land. But Toboso persevered. Even when the Dillon Dam project ripped away the entire lower half of town, Toboso held the high ground.

One summer's day in 1988, a small Ohio town celebrated a bridge. But it may have been a bridge much stronger than concrete and steel. It may have been the bridge that links one heart to another…the bridge that links people to their hometown.

***Carved Brick Design** (right)*

This artistic brick sculpture in front of the Toboso Elementary School depicts the rich history of Black Hand Gorge. An Indian, a canal boat, a steam train in the deep cut, and an interurban car in the tunnel are all depicted.

Carving by Dean Flowers Brick Designs
Photograph by the author, August 1993

Toboso Today

Highway Sign *(above)*
Sign along Route 146 near Toboso.
Photograph by the author, June 1994

Champion Fiddler *(right)*
Kenny Sidle poses with his fiddle, his dog Halfpint, and his '71 Pontiac near the sign that proudly informs visitors that Toboso is Kenny's birthplace. On the hill in the background sits the Toboso Elementary School which Kenny and his 8 brothers and sisters attended. A similar picture was used on the cover of Kenny's 1989 cassette recording entitled, "Fiddle Memories".
Photograph by John Allee Photography
Courtesy of Evelyn Sidle

Bridge Party *(far right)*
The folks of Toboso preparing for their bridge party in 1988.
Photograph by Richard Hewitt, July 1988

Toboso is a quiet little town where everyone seems to know everyone else. It's the kind of place where you don't need to use your turn signals because everybody already knows where you're going. The children of Toboso still walk to the same brick school building that their parents and grandparents walked to in the early years of the century. Some things have changed of course; the canal boats are gone now, the trains don't stop here any more, and the electric interurban is only a fading memory. But the most important part of Toboso is still here – its people.

As you enter Toboso today, you'll see a sign proclaiming it as the birthplace of Kenny Sidle. Kenny's father and uncle started teaching him fiddle tunes in the early 1930's when he was only 4 years old. Kenny has gone on to become one of the best fiddle players in the nation. He has placed in the top ten of the *Grand Masters* competition in Nashville three times, and has claimed more than a dozen state titles. He has been Ohio's champion fiddler five times.

One of Kenny's earliest public performances took place near Toboso on the stage of one of the last "medicine shows" to be held in Licking County. His father made him a special little fiddle which Kenny still has today. He recalls that, when he walked out in front of the crowd that day, he was so nervous that his knees wouldn't stop knocking. But he made it through that performance and hundreds of others since.

The highlight of Kenny's music career (so far) is the *National Heritage Fellowship Award* he received

in Washington D. C. in 1988. Other recipients of this prestigious award include the legendary Doc Watson and Earl Scruggs. Kenny was also awarded the *Distinguished Fiddler Award* at Opryland in 1994. He's had offers to go on the road with some well-known performers, but he always turns them down. He prefers to live a quiet life among family and friends in the small community where he grew up. Kenny, and his wife Evelyn, live in nearby Hanover. He's still fiddlin' around, and folks are still tappin' their toes to his music. Like a fine old fiddle, Kenny just seems to get better with age.

CHAPTER 5

Odds & Ends

A fter researching the history of Black Hand Gorge for several years now, it seems clear that nothing of cosmic significance has ever happened here. No famous battles were fought here, no major discoveries were made in the gorge, no presidents or movie stars were born here, not even any political scandals or alien encounters. The gorge's past isn't very spectacular, it's mostly the history of ordinary people whose beliefs and actions were shaped by the times in which they lived. What makes the gorge's past unique is its geography. Three developing modes of transportation had to squeeze their way through this narrow pass, and they brought with them a parade of people and enterprises that wouldn't normally be so concentrated in a rural setting such as this. The high cliffs of the gorge have always been a natural barrier, resisting man's intrusion. But the boats, trains, and trolleys eventually pushed through anyway, and, in the process, the gorge funneled history into this area.

This chapter offers a few brief glimpses into some interesting enterprises, locations, and curiosities that have come through this funnel over the years, a mix of miscellaneous odds & ends.

***The Cornell Steps** (left)*

Looking like miniature statues, a group of sightseers poses at the Cornell Steps in the early 1900s. Judging by their attire, they may have been taking a ride on the interurban after church. Only the upper flight of steps can be seen clearly here, the lower steps are in the shadow of the large tree.

Photographer Unknown, ca. 1905-10
Collection of Curtis "Bud" Abbott

The Cornell Steps

William and Zabre Cornell owned land a mile or so up-river from Black Hand Rock. Their home was located on a hill overlooking a line of high sandstone cliffs. Just after the turn of the century, the interurban railway laid tracks along the bottom of these cliffs. The Cornell's home was within easy walking distance of the interurban tracks, but there was no way to get down the cliffs which were 60 to 70 feet high. Zabre was interested in getting down to the tracks so she could ride the interurban to Newark to do her shopping. William was interested in keeping Zabre happy.

In 1903, William built Zabre a stairway down the face of the cliff. He didn't make his stairs out of wood or iron, he carved the steps right into the rock. This was no small task. He had to plan carefully so the steps wouldn't be too steep or too narrow. He skillfully carved 54 steps in three flights. From the bottom of the cliff, the first flight goes up to the left. Then there's a small landing and the second flight goes up a few more steps in the same direction. After another landing, the steps change direction and go to the right, all the way to the top of the cliff. We don't know how long it took William to complete his task, but old photographs show that his carving was sharp and consistent. For many years the steps were known as Cornell's Stop on the interurban line.

The Cornell Steps still exist today, but are difficult to see due to the dense overgrowth. The steps are located on private property and should not be approached without permission from the landowner.

Cornell Steps *(top right)*
The first two flights of steps are shown in this old photograph. The young man in the white shirt is Don Hubbard, grandson of William Cornell. The interurban tracks visible at the left side of the picture were eventually removed and replaced by an automobile road. Later still, the road was removed and replaced by the B. & O. Railroad tracks. The tracks are still active today.

Photographer, Bierberg, ca. 1905
Collection of Chance Brockway

Rock Carving *(bottom right)*
This large boulder or outcropping is located along the edge of the railroad tracks near the base of the Cornell Steps. Carved into the rock is a symbol representing stairs and an arrow pointing toward the cliff where the steps are located.

Photograph by the author, April 1994

Postcard *(top right)*

This old picture postcard entitled, "Stone Steps to Country Club, near Newark, Ohio", is somewhat of an enigma. These are clearly the Cornell Steps, but where was the country club? It could be that the postcard was simply printed with an incorrect title, but some people offer a different explanation. They say, that around the turn of the century, there was a "house of ill repute" located on top of this ridge. In polite conversation it was referred to as the "Country Club". We may never know whether this story is true or not, enterprises of this nature were not usually well-documented.

Card by Leighton & Valentine, ca. 1905-10
Collection of the author

Side View *(bottom right)*

This view shows the side of the second landing and the bottom of the upper flight of steps.

Photograph by the author, April 1994

Upper Flight *(left)*

Looking up at the upper flight of steps from the second landing.

Photograph by the author, 1974

The Oil Boom

During the early 1900s, many gas and oil wells were drilled in the area around Black Hand Gorge. For several years the countryside bristled with rigs. A common practice of the time was to "blow out" or "shoot" the oil wells with "nitro" (nitroglycerine). A container of nitro, called a "torpedo", was lowered to the bottom of the well shaft where it was detonated. The explosion cracked the underlying rock formations allowing the oil to flow in easier and collect at the well. Sometimes oil would shoot up out of the well in a tall gusher when the explosion occurred.

Nitroglycerine is a pale yellow liquid that is powerful and extremely volatile. It will explode if dropped or even jarred. To satisfy the need for nitro, manufacturing facilities were built at strategic locations. It was difficult to transport nitro safely, especially in the days when roads and vehicles were not as smooth as today. Around the gorge, nitro was transported to the oil wells in an old Studebaker truck modified for the purpose. It was referred to as the "torpedo truck".

The nitro plants were small, usually consisting of only one building. For obvious reasons, they were located away from populated areas. One of these factories was located at the upper end of the gorge near Hanover. It eventually exploded, seriously injuring one man. Another factory was located near the lower end of the gorge in what is now part of the nature preserve. Around 1928, it exploded too, injuring several workmen and causing considerable property damage. But nitro was still in demand, so a new building was built near the site of the old one. It was a small brick building and was located in a secluded area on the

Shooting an Oil Well *(right)*
This picture was taken in Lima, Ohio around 1910.
Card by Webb Book & Bible Co., ca. 1910
Collection of the author

Wells Near Toboso *(far right)*
This photograph was taken looking north along County Road 273 just northwest of Toboso. The well in the foreground is apparently still in operation today, although it has been replaced by a modern pump.
Photographer Unknown, ca. 1920s
Collection of Goldie Stevens

high ground across the river from Toboso.

Just before noon on March 17, 1930, a tremendous explosion shook the gorge. The blast blew windows out of houses in Toboso and knocked cans and boxes off the shelves of the grocery store. The sound of the explosion was heard for miles in every direction. Everyone new immediately that it had to be the nitro factory. Practically the whole town of Toboso rushed up to the scene of the explosion. Witnesses recall that it looked like a war zone. Thousands of bits and pieces of the building and its contents were scattered across the ground and in the trees. There wasn't much left of the nitro factory. A local man, Willard King, who was working alone at the factory that day, had apparently been in the process of pouring the nitro into a pan or basin when it exploded. He was killed instantly in the blast.

***A Tragic Blast** (right)*

Front page of the Newark Advocate from March 17, 1930.

Courtesy of the Newark Advocate

***Oil Rig** (far right top)*

An old oil rig still standing along Route #146 near Toboso.

Photograph by the author, July 1994

***Old Sign** (far right bottom)*

Jesse Keirns stands beside a sign near the site of the old nitro factory. This tree has grown almost completely around the enameled steel sign. It's impossible to read now, but it probably gave a warning to keep away from the dangerous nitro facility.

Photograph by the author, June 1994

A MILLION years hence, a man may be a really noble creature.—Dean William Ralph Inge.

The Newark Advocate

AMERICAN TRIBUNE

WEATHER—Cloudy and colder tonight and Tuesday with rain tonight and Tuesday.

ADVOCATE ESTABLISHED 1820; AMERICAN 1827. NEWARK, OHIO, MONDAY EVENING, MARCH 17, 1930 VOL. 130, No. 100

MAN KILLED IN NITRO BLAST

PLANS MADE FOR BATTLE AT PRIMARY

Stage Is Set in Pennsylvania for Three-Cornered. Party Fight.

VARE IS BACKING DAVIS

James J. Davis Enters Fight for Control of Republicans.

Pittsburgh, March 17. — (AP) — The stage was set today for an intense three-way fight for control of Pennsylvania Republican politics at the May primaries, with the formal entry of Secretary of Labor James J. Davis into the race for the United States senatorial nomination.

Davis, who has been assured of the backing of William S. Vare and the Philadelphia Republican organization, announced his intention last night of making the race.

Opposing him will be United States Senator Joseph R. Grundy, supported by Governor John S. Fisher, who appointed him to the seat made vacant by the senate's refusal to seat Vare. Fisher appointed Grundy with the specification that Grundy would make the race for the full term.

The third side of the fight for state control will be furnished by the race of former Governor Gifford Pinchot for the Republican gubernatorial nomination. Pinchot will be opposed by Francis Shunk Brown, Philadelphia, who is aligned with Davis, and by Samuel S. Lewis, for-

Second Disastrous Water Front Blaze Hits New Orleans With Big Loss

This spectacular picture shows vividly the onrush of flames and smoke that added another $1,500,000 fire loss to the history of the port of New Orleans. A fire tug is pouring streams of water on the Esplanada avenue warf. It was the second disastrous waterfront fire within two days at New Orleans.

EXPECT COUNCIL TO ACT UPON AIRPORT PROPOSALS TONIGHT

Legislation Is Before City Solons and Final Action Is at Hand; Plan Is for Newark to Give Government Land for Field.

Council action will be taken tonight on the proposal of the United States government to place Newark on a transcontinental air mail route. Legislation necessary for

fered by the city, which has been refused.

It has been pointed out that if Newark is to progress in the future, is to succeed in bringing new industries to this city, is to enlarge the city's population, it will be

WICKERSHAM QUIZZED ON HIS FINDINGS

Senate Committee Is Given First Chance to Find Out About Dry Work.

CONSIDER DRY INQUIRY

House Committee to Resume Hearings on Dry Law Wednesday.

Washington, March 17.—(AP)—A senate committee was given its first opportunity today to question Chairman Wickersham of the Hoover law enforcement commission, on the finding of his group, as they relate to prohibition.

With Attorney General Mitchell, he was called into an executive session of the judiciary committee, which is considering the advisability of undertaking an extensive inquiry into the enforcement of the dry laws and conditions which may be attributed to them.

Today's session dealt with a resolution by the chairman of the committee, Senator Norris of Nebraska, proposing that such an investigation be begun, but the members of the committee were ready with the questions on many phases of the commission's study of the prohibition situation.

It was Wickersham's first appearance before a senate committee since the law enforcement commission was organized, although as a former attorney general, he has testified many times before such groups. He was a witness recently before a committee of the house.

Fastest Liner Built for U. S.

The fastest ship ever built in this country for foreign trade, the liner Santa Clara, is shown above, as it set out on its trial trip off the Delaware Capes. With a speed of 22 knots, the big vessel is expected to clip five days off the schedule from New York to Callao, Peru and Valparaiso. Its maiden voyage to Latin America will begin April 19.

GANGSTER AND WOMAN SLAIN IN CHICAGO

Election Judge Found Beaten to Death, Apparently by Robbers.

JAIL EIGHT HUNDRED

Continue Regular Roundup.

NAVAL GROUP IS REACHING VIEWPOINTS

French and British Find Further Reasons for Ironing-Out Process.

AGREEMENT NOW SEEN

Future Disarmament Conference in Geneva to Have Groundwork.

London, March 17.—(AP)—Some quarters, close to the five-power naval conference, today expressed confidence that Andre Tardieu, French premier, and Prime Minister MacDonald, in a four-hour conversation yesterday, found a ground for further consideration of the differences in their viewpoints, and would now proceed to agreement.

The confidence and optmism were not shared elsewhere. Many felt that unless M. Tardieu made considerable concessions—and it was indicated he did not—the differences between France and Great Britain, and France and the United States, are too essential for ready compromise. They could not see how, other things being what they were, the gulf between the two viewpoints could be bridged.

M. Tardieu himself radiated optimism, but some observers mentioned in this connection that the French viewpoint all along has been this would be a highly successful conference, if a groundwork was laid for a future general disarmament conference at Geneva, and if agreement was attained on technical evaluation of fleets and on humani-

BLAST WRECKS FACTORY UNIT NEAR TOBOSO

L. Willard King of Newark Killed Instantly When Nitro-Glycerine Explosion Occurs at Plant Near Sparse Area at Rock Haven.

L. Willard King, 48, of 26 Boner avenue, manager of the local office of the Producers Torpedo company, Marietta, was instantly killed shortly before noon today when a quantity of nitro-glycerine exploded in a magazine at the Rock Haven plant, two miles from Hanover.

Mr. King was alone at the plant when the explosion occurred.

The explosive fluid was being poured into a catch-all pan, or an overflow basin, and a terrific blast resulted in some unknown manner.

The force of the explosion threw Mr. King a distance of 30 feet, and partially wrecked the nitro-glycerine magazine. Criss Brothers ambulance responded to the scene and removed the body to the King home in Boner avenue.

Mr. King is survived by his wife and two children, a daughter being Mrs. Frank Offenburger, 69 Gainor avenue.

Today's explosion marked the second serious blast at the nitroglycerine plant during the past two years. The first explosion resulted in injuries to several workmen and considerable property damage. The other company employes were not called to the Rock Haven plant this morning and for that reason Mr.

GAS EXPLOSION INJURES TEN

Man Strikes Match in Gas Filled Basement of Apartment House.

Cleveland, March 17. — (AP) — A homeless man seeking a place to sleep, lit a match in a gas-filled apartment house basement last night causing an explosion which injured

Nitro Factory *(above)*

Dewey Lebold (on the left) and Bill Howes, sitting in front of the nitro factory near Toboso. Mr. Howes wore the bandage because of an eye condition.

Photographer Unknown, ca. 1920s
Collection of Curtis "Bud" Abbott

Unidentified Items *(right)*

This glass tube and rod were found at the site of the nitro factory many years after the explosion. When found, the rod was inside the tube.

Photograph by the author, June 1994

The Sand Quarry

Just west of the railroad cut on the south side of the Licking River lies the ruins of the Everett Sand Quarry. The quarry was owned by Edward H. Everett and employed about 50 men. Mr. Everett owned a very successful bottle making factory in Newark. In an 1896 newspaper article, the Edward H. Everett Company was referred to as the largest enterprise in the city of Newark. The main works employed 500 men. The purpose of the quarry at Black Hand Gorge was to produce sand for the manufacture of the glass bottles. The sandstone rock taken from the quarry was processed into a fine, clean sand. The bottles made from the sand were of a superior quality, largely due to the excellence of the sand processed from the Black Hand Sandstone. Thirty to forty tons of glass bottles were manufactured every day at the Newark works.

The quarry was situated on a 50 acre tract at the gorge with direct access to the B&O railroad. Judging from old photographs, it was quite an operation. There were several buildings here as well as a conveyor and at least two wooden towers that look similar to oil derricks. The purpose of the towers is not clear, but they appear to be part of a system of pulleys and cables. Small ore cars on tracks were used to transport the rock around the quarry. There was an incline track that ran to a quarry on top of a large hill. A clever system of cables utilized the weight of the loaded cars coming down the hill to pull the empty cars back up the hill.

The quarry ceased operation many years ago, but the old building foundations and piles of bricks are still visible along the south edge of the bike path. The quarrying pits are filled with water now. The Quarry Rim Trail in the nature preserve passes along the south side of the main quarry and affords an excellent view of this site.

Incline Railroad *(top right)*
The weight of a loaded ore car coming down from the upper quarry, pulls an empty car back up the hill.
Newark Advocate Photo, June 30, 1896
Collection of Curtis "Bud" Abbott

Incline Railroad *(bottom right)*
The tracks are gone now, but the raised rail bed of the old incline railroad is still visible all the way to the upper quarry.
Photograph by the author, November 1994

Quarry View *(far right)*
This photograph was taken from the top of the hill through which the Deep Cut passes. It may have even been taken from atop the wooden tower which was located on the hill. The view is looking southwest, upriver. The B.&.O Railroad tracks running along the left side of the river have now become the bike path. The tracks running along the right side of the river are the electric interurban railway tracks.
Photographer Unknown, ca. 1910
Collection of Chance Brockway

Everett Advertisement *(right)*

This old ad shows an interesting view of the sand quarry looking toward the northeast.

Photographer Unknown, ca. 1880s or 90s
Collection of Curtis "Bud" Abbott

Steam Shovel *(far right)*

Four unidentified quarry workers pose for a picture beside their Marion steam shovel.

Photographer and Date Unknown
Collection of Goldie Stevens

Bottle *(below)*

A bottle made at the Everett Glass Works in Newark.

Collection of Curtis "Bud" Abbott

MARION SHOVEL—MODEL 21
THE
EDWARD H. EVERETT CO
NEWARK, O.
The
Marion
4229

Quarry View *(left)*

Another view looking southwest. Notice the cables coming from the top of the wooden tower at the far side of the quarry and running up off the top of the photograph. These cables were probably attached to the other wooden tower at the opposite side of the quarry. The exact purpose of this cable system is not clear.

Photographer and Date Unknown
Collection of Goldie Stevens

Quarry Workers *(top right)*

Two unidentified gentlemen posing beside the vehicle that was used to pull the ore cars.

Photographer and Date Unknown
Collection of Goldie Stevens

View of Deep Cut *(bottom left)*

A view looking northeast from the interurban tracks across the river from the quarry. The west end of the Deep Cut can be seen near the center of the picture. Notice the end of what appears to be a foot bridge at the bottom right corner.

Photographer Unknown, ca.1905
Collection of Chance Brockway

Loading Cars *(bottom right)*

A steam shovel loading ore cars.

Photographer and Date Unknown
Collection of Goldie Stevens

Cherry Hill Orchard

Around 1905, Edward H. Everett began planting an orchard on a 400 acre tract of land adjacent to his sand quarry at Black Hand Gorge. Mr. Everett owned not only the orchard and the sand quarry, but he also had 14,000 acres in the vicinity of the gorge under lease for drilling gas and oil wells.

The orchard, named Cherry Hill, was located south of the river, just west of Toboso. Like all of Mr. Everett's ventures, the orchard was approached in a scientific and progressive manner. He liked to be the biggest and best in everything he did. Cherry Hill was considered one of the most scientifically advanced orchards in this part of the country.

Mr. Everett hired Herbert A. Albyn to manage Cherry Hill. Albyn planted 3,000 apple trees, 4,200 cherry trees, and 12,000 peach trees. Also planted were pear, quince, and plum trees. A weather station warned Albyn whenever frost was approaching so he could activate his 5,000 smudge-pots to protect the fruit. Rye and clover were sown between the rows of fruit trees to enrich the soil, and twenty stands of honey bees aided in the pollination. If a worker spotted a single tree that might be in trouble, a detailed blueprint of the orchard permitted the foreman to quickly identify the tree and take appropriate action.

The fruits produced at Cherry Hill were of the highest quality. The boys and girls of Toboso sometimes served as unofficial taste-testers of the fruits as they played among the hills near the gorge. Those boys and girls are a little older now but they still rave about the delicious apples and peaches that grew at Cherry Hill Orchard.

As the years went by, Mr. Everett shifted his attention toward his 3600 acre orchard in Vermont, which eventually became America's largest apple orchard, and to his other ventures which included the construction of his mansions in Washington D. C., and Bennington, Vermont.

***Orchard Views** (right)*

Cherry Hill's fruit trees thrived in the hills around Toboso and Black Hand Gorge.

Photographer and Date Unknown
Collection of Curtis "Bud" Abbott

Wickham's Grove

Years ago, there was a beautiful area near the southeast corner of Toboso known as Wickham's Grove. It was a popular gathering spot in the summertime for picnics and family reunions. Veterans of the Civil War held their annual reunions here for several years. An article in the Zanesville Signal of August 10, 1900 reported that: *"The annual reunion of the soldiers and sailors of the civil war was held in Wickham's grove near Black Hand on Thursday and was one of the most successful of this notable series. Between 5,000 and 6,000 people gathered at the assembly grounds to witness or participate in the exercises and to do honor to those veterans for whom no honor is too great".*

The article continued by reporting that the assembly was called to order promptly at 10 a.m. with a musical selection and an opening prayer. After several speeches, the assembly was dismissed for dinner and… *"Tables were spread upon the lawn under the trees and the immense throng resolved itself into innumerable reunions on which the ties were those of family, of friendship or of old associations".* Over 100 gallons of ice cream were consumed that hot summer's day before supplies were exhausted. The afternoon's activities included music and more speeches by prominent citizens from Newark, Coshocton, Clay Lick, and Zanesville, most having titles before their names such as Dr., Rev., or Hon.. The music was provided by the Flint Ridge Band and the Buckeye Band of Newark.

Today, looking out over the empty fields and rolling hills that were once called Wickham's Grove, one can almost imagine what the scene must have been like on that hot August day 95 years ago: Groups of aging veterans reminiscing about the tragic battles they had fought nearly 40 years earlier. Many of them with white beards and canes, some with an arm or leg missing, all with sad stories to tell about places such as Manassas, Vicksburg, Shiloh, and Lookout Mountain. Beneath the trees, ladies in colorful Victorian dresses and feathery hats fan themselves in the shade. Horses and buggies are converging on Toboso from every direction, and maybe even a horseless carriage or two. A train whistle echoes through the gorge announcing another load of visitors arriving at the Toboso station. Long-winded speeches, brass bands playing, children running, and dogs barking – what a spectacle it must have been!

The Zanesville newspaper reporter covering the reunion summed it all up in the final words of his article; *"The Black Hand reunion of 1900 will long be remembered by those present as one of the most pleasurable events of their lives."*

The Holy Stone Controversy

***Decalogue Stone** (below)*
In the top picture, the Decalogue Stone is shown lying in the lower half of the stone box in which it was sealed, the top of the box lies to the left. The bottom picture shows the other side of the Decalogue Stone. The Holy Stones are located at the Johnson-Humrickhouse Museum in Roscoe Village (Coshocton), Ohio.
Photographs by the Johnson-Humrickhouse Museum

In 1860, a man named David Wyrick drew national attention when he dug into Licking County Indian mounds and discovered stones inscribed with Hebrew characters. The stones came to be known as the "Newark Holy Stones". Wyrick, who was somewhat of an eccentric, was especially proficient in mathematics and served as the Licking County Surveyor for several years during the 1850s. To quote a well-known Ohioan of the time, Col. Charles Whittlesey, David Wyrick was: *"wholly a self-taught man, in many respects possessed of genius"*.

The Holy Stones were found at a time when it was theorized that the builders of the mounds were descended from the "Lost Tribes of Israel". It was an emotionally-charged subject with important religious, political, and scientific implications. The only thing lacking was physical proof. Wyrick's stones seemed to prove the Lost Tribes theory at first, but for a variety of reasons, he and the stones were eventually discredited. The stones are now regarded by many scholars as frauds, although the debate still continues.

Since the early 1900s, local books and newspapers have occasionally stated that one of the stones, the Decalogue Stone, was found in a mound at Black Hand Gorge. Over the years, the story has been picked-up and repeated in slightly different versions. One version of the story says Wyrick noticed that the elongated index finger of the Black Hand carving was pointing to a mound across the river. The top of the mound had been removed by the building of the railroad, so he dug into the remaining floor of the mound and found the inscribed stone. Another version says that the Black Hand was pointing toward the Great Stone Mound near Jacksontown, south of Newark.

These stories, and similar versions, have at least one curious thing in common – they link the Black Hand Petroglyph to the Newark Holy Stones. It's an interesting concept. If we assume, for the sake of argument, that the Holy Stones are legitimate, and were made by the people who built the mounds, it's also possible that the Black Hand was a Hebrew symbol. If the Holy Stones are fraudulent, then the connection to the Black Hand is meaningless.

In either case, the stories about the Black Hand pointing Wyrick toward a particular mound are probably fabrications. Wyrick found the Holy Stones in 1860, the Black Hand carving was destroyed in 1828. It's unlikely that David Wyrick ever saw the Black Hand.

The debate about the authenticity of the Newark Holy Stones may never be settled. But there seems to be no real historic connection between the Holy Stones and Black Hand Gorge. Wyrick himself stated that he found the Decalogue Stone at the "...stone mound beyond Jacktown..." (Jacksontown), but makes no mention of the Black Hand pointing the way. Somehow, in later accounts of the discovery, the mysterious Black Hand got into the story. It seems to be human nature to try to link one mystery to another.

The Cow Tunnel

Several yards east of the interurban tunnel, the trail crosses what appears to be a little bridge on the old interurban rail bed. It looks like any ordinary bridge that might be found where small creeks are plentiful. But local folks say it wasn't a typical bridge at all. There is no creek here. It wasn't a stream of water that flowed through this small underpass (or should I say, udderpass) – it was a stream of cows.

Emmet Francis owned the land, and the cows, in this area. He had always watered his cows in the river. The interurban railway coming through would have blocked the cows' pathway to the water, but the dilemma was solved by building a tunnel so the cows could cross safely beneath the interurban tracks.

People around Toboso still remember how Emmet's dairy cows were trained to come to the barn whenever he gave them his special signal. The signal was his car horn. Every evening Emmet would sit in his car and honk his horn 'till the cows came home.

The Cow Tunnel is constructed mostly of cut sandstone blocks. Some of them appear to have been used blocks, probably taken from an old building foundation. Over the years, dirt and debris have washed into the tunnel and accumulated, making the tunnel floor higher than it was originally. Only very short cows could pass through here now.

The inside of the tunnel measures only slightly wider than a cow. According to my cow-culations, an average cow could amble through here without scraping the walls, but I wouldn't want to try to walk beside one through the tunnel. They must have had to enter in single file. This was the Victorian age, an age when everything was done according to proper rules of etiquette. Imagine what might have happened when two cows met in mid-tunnel going in opposite directions. This certainly would have been considered a social blunder and udderly embarrassing for both cows. Loud mooing undoubtedly occurred when this happened and probably continued until Emmet could get out to his car and honk them all home.

Cow Tunnel *(right)*
The little-known Cow Tunnel of Black Hand Gorge.
Photograph by the author, April 1994

The Black Hand Gorge Festival

Festival Scenes, *(both pages)*
Photographs by the author, September 1994

In September 1974, the first "Black Hand Gorge Festival" was held at the Toboso Elementary School. Bill Weaver, the school's principal, gives credit for the festival idea to his long-time friend, Jack Caughenbaugh. Jack thought it would be a good way to help promote the gorge and the local community at the same time. At the core of the festival is a flea market. There were about 30 vendors that first year, but by the 20th annual festival in 1994, the number of vendors had grown to over 300. It has become a major fall event for the community of Toboso. Long tables of items for sale fill the school building, the playground, and the adjoining fields. Hundreds of cars park in the farm fields across from the nature preserve. There is always a parade on festival day. Farm wagons filled with children, horses & carts, a few classic cars, a marching band, and a fire truck or two make their way down the main street in front of the school to the cheers of onlookers. The students of Toboso School prepare a float based on a yearly theme related to the history of the gorge. Cutouts of hands in black construction paper are a favorite decoration for the day. Kenny Sidle and friends usually play their toe-tappin' music in front of the Methodist Church across the street and they always draw quite a crowd.

Food and drink are plentiful at the festival and so are restroom facilities. It's a nice way to spend a Saturday afternoon and a good opportunity to walk in the gorge. Bill Weaver claims that the weather is always perfect on festival day. It's a well-known fact

that Bill is always right about this – well, almost always. The one exception might be the 1994 festival when every rain cloud in the northern hemisphere seemed determined to go out of its way to soak Toboso. The rain came down for hours and hours. Not just a little sprinkle, but a steady downpour. But even the rain couldn't dampen the spirit of the festival. Hundreds of people came anyway and umbrellas began popping open like colorful mushrooms. Vendors sold their goods from under makeshift awnings made out of everything from tarps to trash bags, and the parade marched on in spite of the rain.

There's something for everyone at the Black Hand Gorge Festival. Rows and rows of antiques, arts, crafts, and flea market treasures of all kinds – but some folks come just for the famous homemade noodles!

20th Anniversary *(right)*

These four extremely wet people are (left to right) Jack Caughenbaugh, Cecile Pletcher, Bill Weaver and Marie McClain. They were given Black Hand clocks in recognition of their generous contributions to 20 years of Black Hand Gorge Festivals. The clocks were presented by Mike & Kathy Scott on behalf of the Toboso PTO. Even in the pouring rain, these folks keep on smiling!

Photograph by the author, September 1994

Graffiti

Old Dates *(right)*
This "1809" near the Deep Cut is one of the oldest carved dates in the gorge, dating from the time when Thomas Jefferson was President.
Photographs by the author, May, 1995

Familiar Names *(below)*
Robbins Hunter and Chalmers Pancoast camped in the gorge as boys during the late 1800s. They both carved their names at the base of Black Hand Rock, and they both went on to write books about local history. The date under Pancoast's name reads "97".
Photographs by the author, January 1994

One of the things that seems constant over time, is mankind's penchant for drawing, painting, and carving on the walls of his neighborhood. From the animal drawings in ancient caves to the spray-painted obscenities on today's freeway overpasses, humans seem compelled to send personal messages of all types to everyone else in the vicinity. We tattoo messages on our bodies and our astronauts even left messages on the moon! No blank space is safe when there's a human around with the urge to write or draw. Unfortunately, Black Hand Gorge is no exception.

No matter where you walk at Black Hand Gorge, you are likely to encounter writing or pictures inscribed or painted on the sandstone outcroppings. On the top and bottom of Black Hand Rock, all around the interurban tunnel, in the Deep Cut, and dozens of other places. Some of the markings are old, some more recent – all have done permanent damage.

Graffiti have existed in the gorge as long as man has. The Black Hand petroglyph is a good example. (It's odd that we refer to old carvings, like the Black Hand, as petroglyphs, while we call more recent carvings graffiti. Apparently, if a young graffito survives long enough, it will mature into a petroglyph). These carvings and paintings not only damage the surface of the rock, they also spoil the natural beauty of the surroundings. Some of the more recent graffiti have even been crude and offensive.

The graffiti displayed here are random examples from around the gorge. The intention is not to glorify graffiti by publishing them in a book, but rather to point out that we have more than enough graffiti in the gorge already, and that no more are needed.

High Above the River *(top)*

The top edge of Black Hand Rock, or Council Rock, as it is sometimes called, is covered with graffiti.

Photograph by the author, April 1994

Old and New *(bottom left)*

Along the towpath, the cliff has been defaced by a carving from 1932, which was itself defaced by a spray-painted marijuana leaf in 1994.

Photograph by the author, June 1994

Modern Art? *(bottom right)*

A female figure, a cross, and the ever-popular heart encircling two names, in this case, Dan & Patsy.

Photograph by the author, May 1995

The Nature Preserve

A plaque at the entrance to the Black Hand Gorge State Nature Preserve commemorates the dedication of this site on September 20, 1975. Another plaque was placed nearby in memory of Marie and Lillian Hickey, the sisters who owned the land.

For many years, the gorge was abandoned and unprotected. It became a dumping ground for old tires, trash, even ranges and refrigerators. There have been many people who have given their time and effort over the years to help clean-up and preserve this unique area. Some of the most actively involved have been Bill Weaver, principal of Toboso Elementary School, and his staff and students. In the spring of 1975, Mr. Weaver started a "tour guide" service which trains sixth graders at Toboso as tour guides for students from other schools visiting the gorge. The sixth graders must learn the basics of the gorge's history and geology to become a tour guide. The tour guides lead between 500 and 1000 students through the gorge each year. They come in bus loads from all over Licking and Muskingum counties. The program is still going strong after 20 years, which means several thousand students (many of whom are now adults) have been taken through the gorge where they learned something about its historical significance. Bill Weaver, his staff, and the students of Toboso Elementary School are the unsung heroes of Black Hand Gorge. Their long-term dedication and enthusiasm have helped preserve this unique area for us all to enjoy.

Log Cabin *(far right)*
The old cabin was moved to this location from another area of the gorge.

Main Entrance *(below)*
Preserve sign at the Toboso entrance.
Photographs by the author, June 1994

This is the only state nature preserve in Ohio that has a bike path running through it. The rules of the nature preserve attempt to restrict human intrusion into the natural environment, while at the same time, the presence of a bike path encourages human intrusion. It's a conflict of interests that creates a lot of confusion for visitors. When you visit the nature preserve, pick up a brochure at the entrance. It contains a map showing the various trails and sites as well as a list of rules and regulations. Please remember that this is a nature preserve, not a park. Nature preserves have rules that are much more restrictive than most parks.

Visitors can't wander around wherever they want in the preserve, they must stay on designated trails. In the gorge, it's not always easy to determine what is a trail and what isn't. There are dozens of paths

running through the gorge. Some of these paths intersect with marked trails and are difficult to distinguish from the trails. There are trail signs at the beginning and end of trails, but in between, it's not always apparent where you should or shouldn't walk. In some cases, paths that have been worn smooth by thousands of pairs of feet are, technically, not designated trails and therefore not legal to be on. The preserve ranger will cite you for trespassing if you are seen off the marked trail. If you have a legitimate reason to visit a restricted area of the preserve, apply for a temporary access permit through the preserve ranger or the Ohio Dept. of Natural Resources (ODNR) in Columbus. Some of the photographs in this book show restricted areas of the preserve. Access permits were obtained in order to photograph these areas.

The area of land which comprises the nature preserve is irregular in shape. There were some parcels of land that the ODNR was unable to acquire. The interurban tunnel, for example, is not part of the nature preserve. The same is true for Black Hand Rock and the stone towpath. As mentioned earlier, the Cornell Steps are located on private property and are not part of the preserve. The train trestle and tracks are also private property.

We can only speculate about what the future will hold for Black Hand Gorge. The nature preserve is generally under-staffed and under-funded. Some folks would like to see the area between the train trestle and Toboso Road turned into a park, with picnic areas and possibly a visitor center in the old cabin. Toboso residents have asked that the area around the preserve entrance be mowed regularly, after all, it is also the entrance to their town. Unlike most nature preserves, there's a bike path, canoeing, fishing, hunting, and several historic sites located here. Consequently, the gorge attracts a variety of visitors for a variety of reasons. Facilities for picnicking and a visitor center would probably be welcomed by many of these visitors.

Another question mark in the gorge's future is the huge Longaberger Village™ that has recently broken ground along Route 16, not far from the gorge. According to advance publicity, this complex, covering over 600 acres, will be a major tourist attraction in a few years. It's estimated that the village will employ over 3000 people and will attract millions of visitors. In response to these visitors, hotels, restaurants, gas stations, stores, etc., will be springing up around the area as well. The part of the gorge within the nature preserve is protected from development, but the areas surrounding the preserve are not. It's hard to say what impact this kind of activity will have on the gorge and small towns nearby.

We don't know what the future may bring for this area, but Black Hand Gorge is still a great place to visit, ride bikes, hike, take pictures, and enjoy nature. It's open all year, daylight to dusk – see you there!

Along the Trails *(far right)*
Various scenes along the trails in the preserve: A trail sign, Dogwood blossoms, a tree clinging to a cliff, and visitors, Jesse Keirns and Brian King of Howard, Ohio, hiking along the Canal Lock Trail.

Plaques *(below)*
These plaques are located at the Toboso entrance to the preserve.
Photographs by the author, 1994-95

BLACKHAND ROCK

Miscellaneous

Rope Walker *(right)*
This photograph was made in the gorge sometime in the early 1900s. The rope appears to stretch from the top of Black Hand Rock, where the photographer was probably standing, to either the top of Picnic Rock or across the river to the top of the railroad Deep Cut. The name of the man on the rope is not known.
Photographer Unknown, 1900-1910
Collection of Goldie Stevens

Aerial View *(far right)*
A view of the gorge taken in 1940. The old interurban rail bed was in use as an automobile road at this time. Remnants of the old Cherry Hill Orchard can be seen around the bottom right corner.
ODNR, July 14, 1940

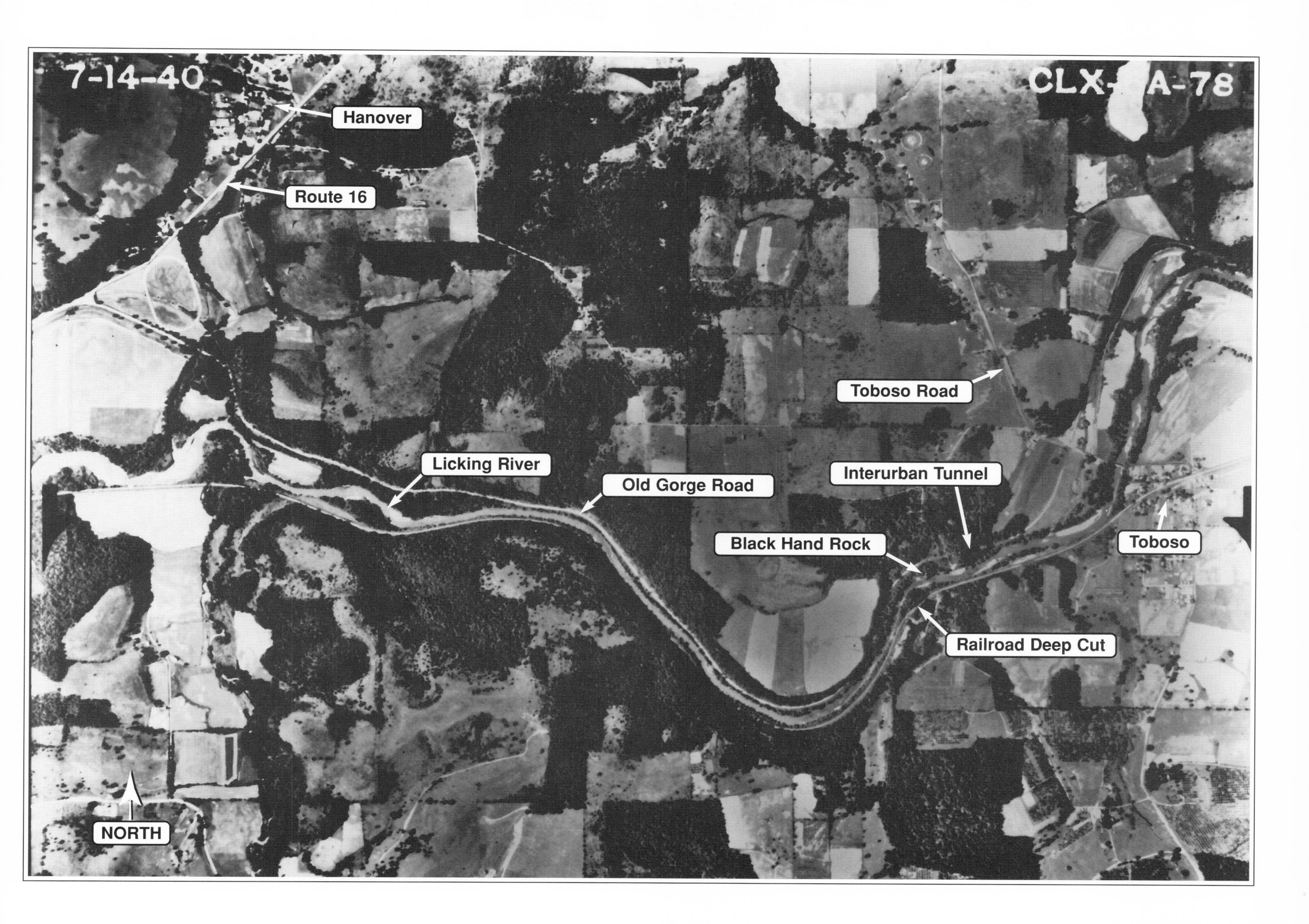

7-14-40
CLX-
A-78
Hanover
Route 16
Toboso Road
Licking River
Old Gorge Road
Interurban Tunnel
Black Hand Rock
Toboso
Railroad Deep Cut
NORTH

Water Fall *(right)*

This scenic waterfall is located northwest of Black Hand Rock. Visitors are not permitted to climb on the falls now, this picture was taken in 1972 before the gorge became a nature preserve. Standing on top of the falls are Bernice and Adam Keirns.

Photograph by the author, 1972

Claylick Cemetery *(far right)*

This old cemetery, near the upper end of the gorge, is all that remains of a small town called Claylick. Claylick once had several homes, a store, school, post office, train station, interurban railway station and, of course, people who called this place "home".

In a letter dated February 5, 1904, a former teacher at Claylick, who had moved away to Johnstown, writes to one of her former Claylick students: "...I miss dear old Newark, and I miss, too, the dearest of all dear places, Clay Lick, and the bright, inspiring faces of its many good boys and girls".

Photograph by the author, May 1995
Old Letter, Courtesy of Greg Hewitt

OF James
Johnson

Then & Now

CHAPTER 6

Then & Now – East End of Tunnel

Photographer, Bierberg ca. 1905
Collection of Chance Brockway

Photograph by the author, May 1995

Then & Now – West End of Tunnel

Photographer Unknown ca. 1910
Collection of Curtis "Bud" Abbott

Photograph by the author, May 1995

Then & Now – East End of Deep Cut

Photographer Unknown, ca. 1900
Collection of Chance Brockway

Photograph by the author, March 1993

Then & Now – West End of Deep Cut

Photographer, Bierberg ca. 1905
Collection of Curtis "Bud" Abbott

Photograph by the author, May 1995

Then & Now – Railroad Culvert Near Quarry

Photographer Unknown ca. 1920
Collection of Goldie Stevens

Photograph by the author, November 1994

Then & Now – Black Hand Dam

Photographer Unknown ca. 1895
Collection of Chance Brockway

Photograph by the author, November 1994

Then & Now – Behind Black Hand Rock

Photographer, Bierberg ca. 1905
Collection of Chance Brockway

Photograph by the author, November 1994

Then & Now – Toboso Bridge

Photographer Unknown, ca. 1920s-30s
Collection of Goldie Stevens

Photograph by the author, November 1994

Then & Now – Cornell Steps

Photographer Unknown ca. 1905-10
Collection of Curtis "Bud" Abbott

Photograph by the author, April 1994

For Further Reading

Alrutz, Robert W.
The Newark Holy Stones, The History of an Archaeological Tragedy
Denison University, 1980

Brister, E. M. P.
Centennial History of the City of Newark and Licking County Ohio *(Vols. 1 & 2)*
The S. J. Clarke Publishing Co., 1909

Chessman, G. Wallace and Curtis W. Abbott
Edward Hamlin Everett, The Bottle King
The Robbins Hunter Museum, 1991

Christiansen, Harry
Ohio Trolley Trails
Transit House, Inc., 1971

Everts, L. H.
1875 History & Atlas, Licking County, Ohio
Adapted from the 1875 Atlas of Licking County
Reprinted by The Bookmark, 1975

Gieck, Jack
A Photo Album of Ohio's Canal Era, 1825-1913
The Kent State University Press, 1988

Hooge, Paul E. and Bradley T. Lepper
Vanishing Heritage
Licking County Archaeology & Landmarks Society, 1992

Howe, Henry
Historical Collections of Ohio *(Vols. 1 & 2)*
State of Ohio, 1900

Hunter, Robbins
The Judge Rode a Sorrel Horse
E. P. Dutton & Co., 1950

M'Kinley, John D. H.
The Black Hand
(Article in) Ohio Archaeological and Historical Publications, Volume XIII, 1904

Pancoast, Chalmers L. and Hazel T. Pancoast
Our Home Town Memories *(Vols. 1 & 2)*
Chalmers L. and Hazel T. Pancoast, 1958

Ramey, Ralph
Fifty Hikes in Ohio
Back Country Publications, 1990

Scott, Harry B. and Karl J. Skutski
The Hanover Story
Braun-Brumfield, 1972

Smucker, Isaac
Centennial History of Licking County, Ohio
Clark & Underwood, Printers, 1876

Smythe, Brandt G.
Mystery of the Black Hand
(Article in) Early Recollections of Newark (Ohio)
Thos. E. Hite Publications, 1940

Swauger, James L.
Petroglyphs of Ohio
Ohio University Press, 1984

Wymer, Dee Anne
Cultural Resource Assessment of the Blackhand Gorge Nature Preserve, Licking County, Ohio
Licking County Archaeology & Landmarks Society, 1986

Index

About the Author

Aaron Keirns grew up in central Ohio and has lived in Columbus, Chesterville, Newark, St. Louisville, Mt. Gilead, Centerburg, and a few places in-between. He received his B. A. in Anthropology from the Ohio State University in 1977, and is employed as a graphic artist and designer. Aaron and his wife, Bernice, have four children: Adam & Tracy who are grown, and Jesse & Nathan, still living at home near Howard, Ohio.

Colophon

This book was designed and produced on an Apple Macintosh IIfx computer. System version 7.1.

Page layout software: *Quark Express*, 3.31.

Illustration software: *Adobe Illustrator*, 5.5.

Image Manipulation: *Adobe Photoshop*, 3.0.

Headlines: *Adobe Garamond*, Bold Italic.

Body Text: *Adobe Garamond*, 12pt

Captions: *Adobe Garamond*, Italic, 9pt.

Files output directly to negative film, printed by offset on Lustro Creme Enamel.

Printed by: Fine Line Graphics, Columbus, Ohio.

Design, layout, and illustration by the author.

Back Cover Scene

Bernice and Adam Keirns about to step into the sunshine from the darkness of the interurban tunnel.

Photograph by the author, 1972